Introductory Statistics
for the Behavioral Sciences

Fifth Edition

Introduction to Statistics
for the Behavioral Sciences

Fifth Edition

Publisher	Earl McPeek
Acquisitions Editor	Lisa Hensley
Market Strategist	Kathleen Sharp
Developmental Editor	Janie Pierce-Bratcher
Project Editor	Michele Tomiak
Art Director	Carol Kincaid
Production Manager	Andrea Archer

Cover image: Controlled Arc #2, 36" x 24", oil paint stick on canvas, © 1998 Neal Parks
Web site: www.ccnet.com/ ~ nparks

ISBN: 0-15-505205-5
Library of Congress Catalog Card Number: **99-61315**

Address for Domestic Orders
Harcourt Brace College Publishers, 6277 Sea Harbor Drive, Orlando, FL 32887-6777
800-782-4479

Address for International Orders
International Customer Service
Harcourt Brace & Company, 6277 Sea Harbor Drive, Orlando, FL 32887-6777
407-345-3800
(fax) 407-345-4060
(e-mail) hbintl@harcourtbrace.com

Address for Editorial Correspondence
Harcourt Brace College Publishers, 301 Commerce Street, Suite 3700, Fort Worth, TX 76102

Web Site Address
http://www.hbcollege.com

Harcourt Brace College Publishers will provide complimentary supplements or supplement packages to those adopters qualified under our adoption policy. Please contact your sales representative to learn how you qualify. If as an adopter or potential user you receive supplements you do not need, please return them to your sales representative or send them to: Attn: Returns Department, Troy Warehouse, 465 South Lincoln Drive, Troy, MO 63379.

Printed in the United States of America

9 0 1 2 3 4 5 6 7 8 039 9 8 7 6 5 4 3 2 1

Harcourt Brace College Publishers

This book is dedicated to
our students—past, present, and future—
to Walter, Julie, Larry, David, Sara, and Ray,
to Judy and Meredith,
and to Pat, Erika, and Gideon.

*This edition is especially dedicated
to Jack Cohen, our mentor and brilliant, loyal friend.
He was loved by so many, including family, friends, and students.*

Preface

This book represents the efforts of three authors who have jointly accumulated many years of experience in statistical procedures through teaching and research. Our purpose has been to introduce and explain statistical concepts and principles clearly and in a highly readable fashion, assuming minimal mathematical sophistication but avoiding a "cookbook" approach to methodology.

We have attempted to present a broader outlook on hypothesis testing than is customary by devoting an entire chapter to the much neglected concepts of statistical power and the probability of a Type II error. To our knowledge, this is the first time that power tables that can be easily used by beginning students of statistics have been included in an introductory statistics textbook. Also included is another important extension of conventional tests of significance: the conversion of t values and other such results of significance tests to correctional-type indicates which express *strength* of relationship. Special time-saving procedures for hand calculation that are outmoded because of the ready availability of electronic calculators and computers, such as the computation of means and standard deviations from grouped frequency distributions, have been omitted.

Throughout the text, the robustness of parametric procedures has been emphasized. However, recognizing the fact that nonparametric tests are widely used, we have included a chapter on this subject. We believe that enough information has been included so that those who use such techniques will be aware of their disadvantages and use them wisely. We have also included a section, within the chapter on analysis of variance, on multiple comparisons. Fisher's LSD method is presented as an extremely useful, though fairly simple comparison method.

In this era of calculators and personal computers, it has become fairly easy to enter data into a machine and obtain the right answer. A considerably more challenging task is that of understanding the rationales that underlie the various statistical procedures, and of interpreting and applying the results without falling prey to common conceptual errors. A major goal of this book, and of the accompanying workbook, has therefore been to emphasize the fundamental logic and proper use of inferential and descriptive statistics.

In this fifth edition, we have emphasized the advantages of using confidence intervals and highlighted some of the disadvantages of null hypothesis testing. While we discuss hypothesis testing in detail because it is so popular in the professional literature, this method also has significant limitations of which students should be aware. In addition, we have taken a more tolerant position regarding one-tailed tests of significance. Finally, the text has been thoroughly reviewed for readability and revised accordingly.

Acknowledgments

Thanks are due to our many encouraging friends and relatives, to colleagues who made many useful comments on our first edition, and most of all to our students who have provided invaluable feedback on both the textbook chapters and the workbook.

We wish to thank Dr. Carol S. Furchner, Dr. Seymour Giniger, and Dr. Joseph A. Trzasko for their most helpful comments and suggestions. We are also indebted to the Literary Executor of the late Sir Ronald A. Fisher, F.R.S., to Dr. Frank Yates, F.R.S., and to Oliver & Boyd, Edinburgh, for permission to reprint Tables C, D, and G from their book *Statistical Tables for Biological, Agricultural, and Medical Research*. Thanks also to the Iowa State University Press; Drs. J. W. Dunlap and A. K. Kurtz; the Houghton Mifflin Company and *Psychometrika;* Dr. A. L. Edwards; the Holt, Rinehart, and Winston Company and the *Annals of Mathematical Statistics* for permission to reproduce the other statistical tables in the Appendix.

Joan Welkowitz
Robert B. Ewen

Contents

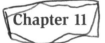

Glossary of Symbols

Numbers in parentheses indicate the chapter in which the symbol first appears.

a_{YX}	Y intercept of linear regression line for predicting Y from X (12)
α	criterion (or level) of significance; probability of Type I error (10)
b_{YX}	slope of linear regression line for predicting Y from X (12)
β	probability of Type II error (10)
$1 - \beta$	power (14)
C	contingency coefficient (17)
cf	cumulative frequency (2)
χ^2	chi square (17)
D	difference between two scores or ranks (11)
$\overline{D}$	mean of the Ds (11)
df	degrees of freedom (10)
df_B	degrees of freedom between groups (15)
df_W	degrees of freedom within groups (15)
df_1	degrees of freedom for factor 1 (16)
df_2	degrees of freedom for factor 2 (16)
$df_{1 \times 2}$	degrees of freedom for interaction (16)
δ	delta (14)
ε	epsilon (15)
f	frequency (2)
f_e	expected frequency (17)
f_m	number of negative difference scores (18)
f_o	observed frequency (17)
f_p	number of positive difference scores (18)
F	statistic following the F distribution (15)

γ	effect size, gamma (14)
G Mdn	grand median (18)
h	interval size (3)
H	statistic following the Kruskal—Wallis test (18)
$H\%$	percent of subjects in all intervals higher than the critical one (3)
H_0	null hypothesis (10)
H_1	alternative hypothesis (10)
i	case number (1)
$I\%$	percent of subjects in the critical interval (3)
k	a constant (1)
k	number of groups (or the last group) (15)
$L\%$	percent of subjects in all intervals below the critical one (3)
LRL	lower real limit (3)
LSD	Fisher protected t test (15)
Mdn	median (4)
MS	mean square (15)
MS_B	mean square between groups (15)
MS_W	mean square within groups (15)
MS_1	mean square for factor 1 (16)
MS_2	mean square for factor 2 (16)
$MS_{1\times2}$	mean square for interaction (16)
μ	population mean (4)
N	number of subjects or observations (1)
N_G	number of observations or subjects in group G (15)
π	hypothetical population proportion (10)
ρ	observed sample proportion (10)
$P(A)$	probability of event A (8)
PR	percentile rank (3)
ϕ	phi coefficient (17)
r_C	matched pairs rank biserial correlation coefficient (18)
r_G	Glass rank biserial correlation coefficient (18)
r_{pb}	point-biserial correlation coefficient (13)
r_s	Spearman rank-order correlation coefficient (13)
r_{XY}	sample Pearson correlation coefficient between X and Y (12)
$\overline{R}$	mean of a set of ranks (18)
ρ_{XY}	population correlation coefficient between X and Y (12)

s	sample standard deviation (5)
s^2	population variance estimate (5)
S_D^2	variance of the Ds (11)
s_{pooled}^2	pooled variance (11)
$s_{\bar{X}}$	standard error of the mean (10)
$s_{\bar{X}_1 - \bar{X}_2}$	standard error of the difference (11)
$s_{Y'}$	estimate of σ_γ obtained from a sample (12)
$Score_p$	score corresponding to the pth percentile (3)
SFB	sum of frequencies below the critical interval (3)
SS	sum of squares (15)
SS_T	total sum of squares (15)
SS_B	sum of squares between groups (15)
SS_W	sum of squares within groups (15)
SS_1	sum of squares for factor 1 (16)
SS_2	sum of squares for factor 2 (16)
$SS_{1\times 2}$	sum of squares for interaction (16)
Σ	sum or add up (1)
σ	standard deviation (5)
σ^2	variance (5)
σ_p	standard error of a sample proportion (10)
σ_T	standard error of the ranks of independent samples (18)
σ_{T_M}	standard error of the ranks of matched samples (18)
$\sigma_{\bar{X}}$	standard error of the mean when σ is known (10)
$\sigma_{Y.}$	standard error of estimate for predicting Y (12)
t	statistic following the t distribution (10)
T	T score (6)
T_E	expected sum of the ranks (18)
T_i	sum of ranks in group i (18)
θ	theta (14)
x	deviation score (4)
X'	predicted X score (12)
$\bar{X}$	sample mean (4)
$\bar{X}_G$	mean of group G (15)
Y'	predicted Y score (12)
z	standard score based on a normal distribution (9)
Z	standard score (6)

Introductory Statistics
for the Behavioral Sciences

Fifth Edition

Part I
Introduction

Chapter 1   Introduction

Chapter 1
Introduction

PREVIEW

Why Study Statistics?

What are three important reasons why a knowledge of statistics is essential for anyone majoring in psychology, sociology, or education?

Descriptive and Inferential Statistics

What is the difference between descriptive and inferential statistics? Why must behavioral science researchers use inferential statistics?

Populations, Samples, Parameters, and Statistics

What is the difference between a population and a sample?

Why is it important to specify clearly the population from which a sample is drawn?

What is the difference between a parameter and a statistic?

Summation Notation

Why is summation notation used by statisticians?

What are the eight rules involving summation notation?

Summary

Why Study Statistics?

This book is written primarily for undergraduates majoring in psychology, sociology, and education. There are three reasons why a knowledge of statistics is essential for those who wish to pursue the study of these behavioral sciences:

To understand the professional literature. Most professional literature in the behavioral sciences includes results that are based on statistical analyses. Therefore, you will be unable to understand important articles in scientific journals and books unless you understand statistics. It is possible to seek out secondhand reports that have been carefully edited in order to be comprehensible to the statistically ignorant, but those who prefer this alternative to obtaining first-hand information should not be majoring in the fields of behavioral science.

To understand the rationale underlying research in the behavioral sciences. Statistics is not just a catalog of procedures and formulas. It offers the rationale upon which much of behavioral science research is based — namely, drawing inferences about a population based on data obtained from a sample. Those familiar with statistics understand that research consists of a series of educated guesses and fallible decisions, and not right or wrong answers. Those without a knowledge of statistics, on the other hand, cannot truly understand the strengths and weaknesses of the techniques used by behavioral scientists to collect information and draw conclusions.

To carry out behavioral science research. In order to do competent research in the behavioral sciences, it is necessary to design the statistical analysis *before* the data are collected. Otherwise, the research procedures may be so poorly planned that not even an expert statistician can make any sense out of the results. To be sure, it is possible (and often advisable) to consult someone more experienced in statistics for assistance. Without some statistical knowledge of your own, however, you will find it difficult or impossible to convey your needs to someone else and to understand the replies.

Save for these introductory remarks, we do not regard it as our task to persuade you that statistics is important in psychology, sociology, and education. If you are seriously interested in any of these fields, you will find this out for yourself. Accordingly, this book contains no documented examples selected from the professional literature to prove to you that statistics really is used in these fields. Instead, we have used realistic examples with numerical values chosen so as to reveal the issues involved as clearly as possible.

We have tried to avoid a "cookbook" approach that places excessive emphasis on computational recipes. The various statistical procedures and the essential underlying concepts have been explained at length, and insofar as possible in standard English, so that you will know not only what to do but why you are doing it. Do not, however, expect to learn the material in this book from a single reading. The concepts involved in statistics, especially inferential statistics, are so challenging that it is often said that the only way to completely understand statistics is to teach it (or write a book about it). On the other hand, there is no reason to approach statistics with fear and trembling. You do not have to be a mathematical expert to obtain a good working knowledge of statistics. What *is* needed is mathematical comprehension sufficient to cope with high-school algebra and a willingness to work at new concepts until they are understood.

Descriptive and Inferential Statistics

One purpose of statistics is to summarize or describe the characteristics of a set of data in a clear and convenient fashion. This is accomplished by what are called *descriptive statistics.* For example, your grade point average serves as a convenient summary of all of the grades that you have received in college. Part 2 of this book is devoted to descriptive statistics.

A second function of statistics is to make possible the solution of an extremely important problem. Behavioral scientists can never measure *all* of the cases in which they are interested. For example, a clinical psychologist studying the effects of various kinds of therapies cannot obtain data on all of the mental patients in the world; a social psychologist studying ethnic differences in attitudes cannot measure all of the millions of Blacks and Whites in the United States; an experimental psychologist cannot observe the maze behavior of all rats. Behavioral scientists want to know what is happening in a given *population*—a large group (theoretically an infinitely large group) of people, animals, objects, or responses that are alike in at least one respect (for example, all Blacks in the United States). They cannot measure the entire population, however, because it is so large that it would be much too time consuming and expensive to do so. What to do?

One reasonable procedure is to measure a relatively small number of cases drawn from the population (that is, a *sample*). A sample of 100 people can readily be interviewed or given a written questionnaire. However, conclusions that apply only to the 100 people who happened to be included in the sample are unlikely to be of much interest. The behavioral scientist hopes to advance scientific knowledge by drawing general conclusions—for example, about the populations of Blacks and Whites from which the samples of 50 Blacks and 50 Whites were drawn. *Inferential statistics* makes it possible to draw inferences about what is happening in the population based on what is observed in a sample from that population. (This point will be discussed at greater length in Chapter 8.) Part 3 of this book is devoted to inferential statistics, which make frequent use of some of the descriptive statistics discussed in Part 2.

Populations, Samples, Parameters, and Statistics

As the above discussion indicates, the term *population* as used in statistics does not necessarily refer to people. For example, the population of interest might be that of all white rats of a given genetic strain, or all responses of a single subject's eyelid in a conditioning experiment.

Whereas the population consists of all of the cases of interest, a *sample* consists of any subgroup drawn from the specified population. It is important that the population be clearly specified. For example, a group of 100 New York University freshmen might be a well-drawn sample from the population of all NYU freshmen or a poorly drawn sample from the population of all undergraduates in the United States. It is strictly proper to apply (that is, *generalize*) the research results only to the specified population. (A researcher *may* justifiably argue that her results are more widely generalizable, but she is on her own if she does so because the rules of statistical inference do not justify this.)

A *statistic* is a numerical quantity (such as an average) that summarizes some characteristic of a sample. A *parameter* is the corresponding value of that characteristic in the population. For example, if the average studying time of a sample of 100 NYU freshmen is 7.4 hours per week, then 7.4 is a statistic. If the average studying time of the population of all NYU freshmen is 9.6 hours per week, then 9.6 is the corresponding population parameter. Usually the values of population parameters are unknown because the population is too large to measure in its entirety, and appropriate techniques of inferential statistics are used to estimate the values of population parameters from sample statistics. If the sample is properly selected, the sample statistics will often give good estimates of the parameters of the population from which the sample was drawn; if the sample is poorly chosen, erroneous conclusions are likely to occur. Whether you are doing your own research or reading about that of someone else, you should always check to be sure that the population to which the results are generalized is proper in light of the sample from which the results were obtained.

Summation Notation

Mathematical formulas and symbols often appear forbidding. In fact, they save time and trouble by clearly and concisely conveying information that would be awkward to express in words. In statistics, a particularly important symbol is the one used to represent the *sum* of a set of numbers (the value obtained by adding up the numbers).

To illustrate the use of summation notation, let us suppose that eight students take a ten-point quiz. Letting X stand for the variable in question (quiz scores), let us further suppose that the results are as follows:

$$X_1 = 7 \quad X_2 = 9 \quad X_3 = 6 \quad X_4 = 10$$
$$X_5 = 6 \quad X_6 = 5 \quad X_7 = 3 \quad X_8 = X_N = 4$$

Notice that X_1 represents the first score on X; X_2 stands for the second score on X; and so on. Also, the *number of scores* is denoted by N; in this example, $N = 8$. The last score may be represented either by X_8 or X_N. The *sum of all the X scores* is represented by

$$\sum_{i=1}^{N} X_i$$

where $\sum$, the Greek capital letter sigma, stands for "the sum of" and is called the *summation sign*. The subscript below the summation sign indicates that the sum begins with the first score (X_i where $i = 1$), and the superscript above the summation sign indicates that the sum continues up to and including the last score (X_i where $i = N$ or 8). Thus,

$$\sum_{i=1}^{N} X_i = X_1 + X_2 + X_3 + X_4 + X_5 + X_6 + X_7 + X_8$$

$$= 7 + 9 + 6 + 10 + 6 + 5 + 3 + 4$$

$$= 50$$

In some instances, the sum of only a subgroup of the numbers may be needed. For example, the symbol

$$\sum_{i=3}^{6} X_i$$

represents the sum beginning with the third score (X_i where $i = 3$) and ending with the sixth score (X_i where $i = 6$). Thus,

$$\sum_{i=3}^{6} X_i = X_3 + X_4 + X_5 + X_6$$

$$= 6 + 10 + 6 + 5$$

$$= 27$$

Most of the time, however, the sum of *all* the scores is needed in the statistical analysis. In such situations it is customary to omit the indices i and N from the notation, as follows:

$$\sum X = \text{sum of all the } X \text{ scores}$$

The fact that there is no written indication as to where to begin and end the summation is taken to mean that all the X scores are to be summed.

Summation Rules

Certain rules involving summation notation will prove useful in subsequent chapters. Let us suppose that the eight students previously mentioned take a second quiz, denoted by Y. The results of both quizzes can be summarized conveniently as follows:

Subject (S)	Quiz 1 (X)	Quiz 2 (Y)
1	7	8
2	9	6
3	6	4
4	10	10
5	6	5
6	5	10
7	3	9
8	4	8

We have already seen that $\sum X = 50$. The sum of the scores on the second quiz is equal to:

$$\sum Y = Y_1 + Y_2 + Y_3 + Y_4 + Y_5 + Y_6 + Y_7 + Y_8$$

$$= 8 + 6 + 4 + 10 + 5 + 10 + 9 + 8$$

$$= 60$$

The following rules are illustrated using this small set of data shown, and you should verify each one carefully.

Rule 1. $\sum (X + Y) = \sum X + \sum Y$

Illustration S	X	Y	X + Y
1	7	8	15
2	9	6	15
3	6	4	10
4	10	10	20
5	6	5	11
6	5	10	15
7	3	9	12
8	4	8	12
	$\sum X = 50$	$\sum Y = 60$	$\sum (X + Y) = 110$
		$\sum X + \sum Y = 110$	

This rule should be intuitively obvious; the same total should be reached regardless of the order in which the scores are added.

Rule 2. $\sum (X - Y) = \sum X - \sum Y$

Illustration S	X	Y	X − Y
1	7	8	−1
2	9	6	3
3	6	4	2
4	10	10	0
5	6	5	1
6	5	10	−5
7	3	9	−6
8	4	8	−4
	$\sum X = 50$	$\sum Y = 60$	$\sum (X - Y) = -10$
		$\sum X - \sum Y = -10$	

Similar to the first rule, it makes no difference whether you subtract first and then sum $[\sum (X - Y)]$ or obtain the sums of X and Y first and then subtract $(\sum X - \sum Y)$.

Unfortunately, matters are not so simple when multiplication and squaring are involved.

Rule 3. $\sum XY \neq \sum X \sum Y$

That is, first multiplying each X score by the corresponding Y score and then summing ($\sum XY$) is *not* equal to summing the X scores ($\sum X$) and summing the Y scores ($\sum Y$) first and then multiplying once ($\sum X \sum Y$).

Illustration S	X	Y	XY
1	7	8	56
2	9	6	54
3	6	4	24
4	10	10	100
5	6	5	30
6	5	10	50
7	3	9	27
8	4	8	32
	$\sum X = 50$	$\sum Y = 60$	$\sum XY = 373$

$$\sum X \sum Y = (50)(60) = 3,000$$

Observe that $\sum XY = 373$, while $\sum X \sum Y = 3,000$.

Rule 4. $\sum X^2 \neq (\sum X)^2$

That is, first squaring all of the X values and then summing ($\sum X^2$) is *not* equal to summing first and then squaring a single quantity [$(\sum X)^2$].

Illustration S	X	X^2
1	7	49
2	9	81
3	6	36
4	10	100
5	6	36
6	5	25
7	3	9
8	4	16
	$\sum X = 50$	$\sum X^2 = 352$

$$(\sum X)^2 = (50)^2 = 2,500$$

Here, $\sum X^2 = 352$, while $(\sum X)^2 = 2,500$.

Rule 5.   If k is a *constant* (a fixed numerical value), then

$$\sum k = Nk$$

Illustration S	Suppose that $k = 3$. Then, k
1	3
2	3
3	3
4	3
5	3
6	3
7	3
8	3
	$\sum k = 24$
	$Nk = (8)(3) = 24$

Rule 6.   If k is a constant,

$$\sum (X + k) = \sum X + \sum k = \sum X + Nk$$

	Illustration	Suppose that $k = 5$. Then,	
S	X	k	X + k
1	7	5	12
2	9	5	14
3	6	5	11
4	10	5	15
5	6	5	11
6	5	5	10
7	3	5	8
8	4	5	9
	$\sum X = 50$	$\sum k = Nk = 40$	$\sum (X + k) = 90$
	$\sum X + Nk = 50 + (8)(5) = 90$		

This rule follows directly from Rules 1 and 5.

Rule 7.   If k is a constant,

$$\sum (X - k) = \sum X - Nk$$

The illustration of this rule is similar to that of Rule 6 and is left to the reader as an exercise.

Rule 8. If k is a constant,

$$\sum kX = k \sum X$$

S	Illustration Suppose that $k = 2$. Then,		
	X	k	kX
1	7	2	14
2	9	2	18
3	6	2	12
4	10	2	20
5	6	2	12
6	5	2	10
7	3	2	6
8	4	2	8
	$\sum X = 50$		$\sum kX = 100$

$$k \sum X = (2)(50) = 100$$

Summary

Descriptive statistics are used to summarize and make understandable large quantities of data. *Inferential statistics* are used to draw inferences about numerical quantities (called *parameters*) concerning *populations* based on numerical quantities (called *statistics*) obtained from *samples*. The summation sign, $\sum$, is used to indicate "the sum of" and occurs frequently in statistical work. Remember that $\sum X$ is a shorthand version of

$$\sum_{i=1}^{N} X_i$$

(N = number of subjects or cases).

1. $\sum (X + Y) = \sum X + \sum Y$
2. $\sum (X - Y) = \sum X - \sum Y$
3. $\sum XY$ (multiply first, then add) $\neq \sum X \sum Y$ (add first, then multiply)
4. $\sum X^2$ (square first, then add) $\neq (\sum X)^2$ (add first, then square)

If k is a constant,

5. $\displaystyle\sum k = Nk$
6. $\displaystyle\sum (X + k) = \sum X + Nk$
7. $\displaystyle\sum (X - k) = \sum X - Nk$
8. $\displaystyle\sum kX = k \sum X$

Part II
Descriptive Statistics

Chapter 2
Frequency Distributions and Graphs

PREVIEW

The Purpose of Descriptive Statistics

What is the primary purpose of descriptive statistics?

What are the four important varieties of descriptive statistics?

Regular Frequency Distributions

What is a regular frequency distribution, how is it constructed, and why is it useful?

Cumulative Frequency Distributions

How does a cumulative frequency distribution differ from a regular frequency distribution, and how is it constructed?

Grouped Frequency Distributions

How does a grouped frequency distribution differ from a regular frequency distribution, and how is it constructed?

What is gained by using a grouped frequency distribution?

What is lost by using a grouped frequency distribution, and why are such distributions usually *not* used when computing means and other statistics?

Graphic Representations

What are histograms, stem-and-leaf displays, and frequency polygons?

When is a histogram preferable to a frequency polygon?

When is a frequency polygon preferable to a histogram?

Shapes of Frequency Distributions

What is meant when we say that a distribution is symmetric? skewed? unimodal? bimodal? normal? rectangular? a J-curve?

Summary

The Purpose of Descriptive Statistics

The primary goal of descriptive statistics is to bring order out of chaos. For example, consider the plight of a professor who has given an examination to a class of 85 students and has computed the total score on each student's exam. In order to decide what represents good and bad performance on the examination, the professor must find a way to comprehend and interpret 85 numbers (test scores). Similarly, a researcher who runs 60 rats through a maze and records the time taken to run the maze on each trial is faced with the problem of interpreting 60 numbers.

In addition to causing problems for the professor or researcher, the large quantity of numbers also creates difficulties for the audience in question. The students in the first example are likely to request the distribution of test scores so as to be able to interpret their own performance, and they also will have trouble trying to interpret 85 unorganized numbers. Likewise, the people who read the scientific paper ultimately published by the researcher interested in rats and mazes will have a difficult time trying to interpret a table with 60 numbers in it.

Descriptive statistics help to resolve problems such as these by making it possible to *summarize and describe large quantities of data.* Among the various techniques that you will find particularly useful are the following:

Frequency distributions and graphs—procedures for describing all (or nearly all) of the data in a convenient way.

Measures of "central tendency"—single numbers that describe the location of a distribution of scores: where it generally falls within the infinite range of possible values.

Measures of variability—single numbers that describe how "spread out" a set of scores is: whether the numbers are similar to each other and vary very little, as opposed to whether they tend to be very different from one another and vary a great deal.

Transformed scores—new scores that replace each original number, and show at a glance how good or bad any score is in comparison to the other scores in the group.

Each of these procedures serves a different (and important) function. Our discussion of descriptive statistics will begin with frequency distributions and graphs; the other topics will be treated in subsequent chapters.

Regular Frequency Distributions

One way of making a set of data more comprehensible is to write down every possible score value in order, and next to each score value record the number of times that the score occurs. For example, suppose that an experimenter interested in human ability obtains a group of 100 college undergraduates and gives each one 10 problems to solve. A subject's score

TABLE 2.1

Number of problems solved correctly by 100 college
undergraduates (hypothetical data)

5	8	3	6	5	8	3	7	4	7
6	8	7	4	6	5	8	7	10	6
5	3	6	1	8	6	8	5	7	10
8	9	6	9	6	5	9	9	6	3
8	5	8	4	8	6	7	4	10	5
8	7	6	8	5	7	6	9	6	10
4	8	6	8	6	5	4	7	9	6
7	4	5	5	9	6	6	7	4	5
10	3	5	7	9	10	6	7	6	6
6	9	8	7	8	5	7	4	6	8

may therefore fall between zero (none correct) and 10 (all correct), inclusive. The scores of the 100 subjects are shown in Table 2.1.

As you can see, the table of 100 numbers is difficult to interpret. A *regular frequency distribution*, on the other hand, will present a clearer picture. The first step in constructing such a distribution is to list every *score value* in the first column of a table (frequently denoted by the symbol X), with the highest score at the top. The *frequency* (denoted by the symbol f) of each score, or the number of times a given score was obtained, is listed to the right of the score in the second column of the table. To arrive at the figures in the "frequency" column, you could go through the data and count all the tens, go through the data again and count all the nines, and so forth, until all frequencies were tabulated. A more efficient plan is to go through the data just once and make a tally mark next to the appropriate score in the score column for each score, and add up the tally marks at the end.

The complete regular frequency distribution is shown in Table 2.2 (ignore the "cumulative frequency" column for the moment; it will be

TABLE 2.2

Regular and cumulative frequency distributions
for data in Table 2.1

Score (X)	Frequency (f)	Cumulative Frequency (cf)
10	6	100
9	9	94
8	17	85
7	15	68
6	23	53
5	15	30
4	9	15
3	5	6
2	0	1
1	1	1
0	0	0

discussed in the next section). The table reveals at a glance how often each
score was obtained. For example, nine people received a score of 9 and five
people received a score of 3. This makes it easier to interpret the performance of the subjects in this experiment, since you can conveniently ascertain (among other things) that 6 was the most frequently obtained score,
scores distant from 6 tended to occur less frequently than scores close to 6,
and the majority of people got more than half the problems correct.

Cumulative Frequency Distributions

The primary value of *cumulative frequency distributions* will not become apparent until subsequent chapters, when they will prove to be of assistance in
the computation of certain statistics (such as the median and percentiles).
To construct a cumulative frequency distribution, first form a regular frequency distribution. Then, start with the *lowest* score in the distribution and
form a new column of *cumulative frequencies* by adding up the frequencies
as you go along. For example, the following diagram shows how the cumulative frequencies (denoted by the symbol *cf*) in the lower portion of Table 2.2
were obtained:

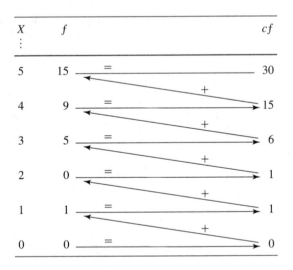

The cumulative frequency for the score of 4 is equal to 15. This value
was obtained by adding the frequency (*f*) of the next lower score, namely 9,
to the cumulative frequency (*cf*) of the next lower score, which is 6. The
other cumulative frequencies were obtained in a similar fashion.

The cumulative frequency distribution is interpreted as follows: the cumulative frequency of 15 for the score of 4 means that 15 people obtained a
score of *4 or less*. This can be readily verified by looking at the frequencies

(*f*): nine people scored 4, five people scored 3, and one person scored 1, for a total of 15 people with scores at or below 4:

X	f		cf	
⋮				
5	15		30	
4	9	scores of 4	15	scores of 4 or less
3	5	scores of 3	6	
2	0	scores of 2	1	
1	1	scores of 1	1	
0	0	score of 0	0	

The *cf* for the score of 7 in Table 2.2 is 68, which means that 68 people obtained a score of *7 or less*. Since everyone obtained a score of *10 or less*, the *cf* for a score of 10 equals 100, the total number of subjects. (The *cf* for the *highest* score should always equal N.) Now, can you tell how many people obtained a score of *6 or more?* The cumulative frequency distribution reveals that 30 people obtained scores of 5 or less, so there must be *70* people who scored 6 or more. This can be checked by using the frequencies; there were *23* scores of 6 + *15* scores of 7 + *17* scores of 8 + *9* scores of 9 + *6* scores of 10. Although no unique information is presented by the cumulative frequency distribution, it allows you to arrive at certain needed information more quickly and conveniently than is possible using the regular frequency distribution.

Grouped Frequency Distributions

When the number of *different scores* to be listed in the score (*X*) column is not too large, regular frequency distributions are an excellent way to summarize a set of data. If, however, there are more than 15 or 20 values of *X* to be written, constructing a regular frequency distribution is likely to prove tedious. One way of avoiding an excessive case of writer's cramp, and a summary that does not summarize enough, is to use a *grouped frequency distribution*. Instead of listing single scores in the score column (for example, 0, 1, 2, 3, 4, 5, 6, 7, . . .), several score values are grouped together into a *class interval* (for example, 0–4, 5–9, 10–14, 15–19, . . .), and frequencies are tallied for each interval.

For example, the data in Table 2.3 represent hypothetical scores of 85 students on a 50-point midterm examination. Inspection of the data shows

TABLE 2.3

Scores of 85 students on a 50-point midterm examination
(hypothetical data)

39	42	30	11	35	25	18	26	37	15
29	22	33	32	21	43	11	11	32	29
44	26	30	49	13	38	26	30	45	21
31	28	14	35	10	41	15	39	33	34
46	21	38	26	26	37	37	14	26	24
32	15	22	28	33	47	9	22	31	20
37	40	20	39	30	18	29	35	41	21
26	25	29	33	23	30	43	28	32	32
34	28	38	32	31					

that the scores range from a low of 9 to a high of 49. Therefore, if a regular frequency distribution were to be used, some 41 separate scores and corresponding frequencies would have to be listed.* To avoid such a tiresome task, a grouped frequency distribution has been formed in Table 2.4. An *interval size* of 3 has been chosen, meaning that there are three score values in each class interval. (The symbol h will be used to denote interval size.) Then, successive intervals of size 3 are formed until the entire range of scores has been covered. Next, frequencies are tabulated, with all scores falling in the same interval being treated equally. For example, a score of 39, 40, or 41 would be entered by registering a tally mark next to the class interval 39–41. When the tabulation is completed, the frequency opposite

TABLE 2.4

Grouped and cumulative frequency distributions for data in
Table 2.3

Class Interval	Frequency (f)	Cumulative Frequency (cf)
48–50	1	85
45–47	3	84
42–44	4	81
39–41	6	77
36–38	7	71
33–35	9	64
30–32	14	55
27–29	8	41
24–26	10	33
21–23	8	23
18–20	4	15
15–17	3	11
12–14	3	8
9–11	5	5

* There are 49 − 9 + 1 or 41 numbers between 9 and 49, inclusive.

a given class interval indicates the number of cases with scores in that interval.

Note that grouped frequency distributions lose information, since they do not provide the exact value of each score. They are very convenient for purposes of summarizing a set of data, but should not generally be used when computing means and other statistics.

In the illustrative problem, the interval size of 3 was specified. In your own work, there will be no such instructions, and it will be up to you to construct the proper intervals. The conventional procedure is to select the intervals in such a way as to satisfy the following guidelines:

1. Have a total of approximately 8 to 15 class intervals.

2. Use an interval size of 2, 3, 5, or a multiple of 5, selecting the smallest size that will satisfy the first rule. (All intervals should be the same size.)

3. Make the lowest score in each interval a multiple of the interval size.

For example, suppose that scores range from a low of 49 to a high of 68. There is a total of 20 score values $(68 - 49 + 1)$, and an interval size of 2 will yield 20/2 or 10 class intervals. This falls within the recommended limits of 8 and 15, and size 2 should therefore be selected. It would waste too much time and effort to list all the nonoccurring scores between zero and 48. Make the first interval 48–49, the next interval 50–51, and so on, so that the first score in each interval will be evenly divisible by the interval size, 2.

If instead scores range from a low of zero to a high of 52, there are 53 score values in all. An interval size of 2 or 3 will yield too many intervals $(53/2 = 26+; 53/3 = 17+)$, and size 4 is customarily avoided (being so close to 5, which produces intervals more like the familiar decimal system). Therefore, interval size 5, which will produce 11 intervals, should be chosen. Begin with the class interval 0–4 and continue with 5–9, 10–14, 15–19, and so on.

What if scores range from 14 to 25? In this case, there are only 12 possible score values and you *should not group* the data, because not enough time and effort will be saved to warrant the loss of information. Use a regular frequency distribution.

Just as was the case with regular frequency distributions, a cumulative frequency distribution can be formed from grouped data, and you will find one in Table 2.4. The cumulative frequencies are formed by starting with the lowest interval and adding up the frequencies as you go along, and are interpreted in the usual way. For example, the value of 64 corresponding to the class interval 33–35 means that 64 people obtained scores at or below this interval—that is, scores of 35 or less.

Graphic Representations

It is often effective to express frequency distributions pictorially as well as in tables. Three procedures for accomplishing this are discussed next.

Histograms

Suppose that a sample of 25 families is obtained and the number of children in each family is recorded, and the results are as follows:

Number of Children (X)	f
7 or more	0
6	1
5	0
4	3
3	4
2	8
1	5
0	4

A *histogram* (or "bar graph") of these data is shown in Figure 2.1. To construct the histogram, the Y axis (vertical axis) is marked off in terms of *frequencies,* and the X axis (horizontal axis) is marked off in terms of *score values.* The frequency of any score is expressed by the height of the bar

FIGURE 2.1

Histogram expressing number of children per family in a sample of 25 families (hypothetical data)

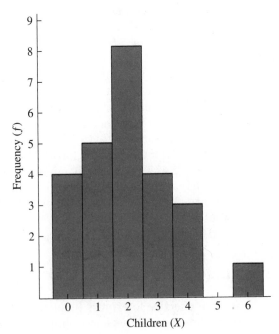

above that score. For example, to show that a score of 3 (children) occurred 4 times, a bar 4 units in height is drawn above this score. To construct a histogram from a grouped frequency distribution, simply use the midpoint of each class interval (the average of the highest and lowest scores in the interval)* as the score value for that interval.

Histograms can be used for any kind of data but are particularly appropriate for *discrete* data, where results between the score values shown *cannot* occur. In the present example, it is impossible for a family to have 2.4 children, and this fact is well expressed in the histogram by the separate and distinct bars above each score value.

Stem-and-Leaf Displays

A simple and useful technique for summarizing a set of data, the *stem-and-leaf display* is a hybrid that combines features of the frequency distribution and the histogram.† The "stems" consist of class intervals, while the "leaves" are strings of specific values within each interval.

To illustrate, consider once again the scores of the 85 students shown in Table 2.3 (and presented as a grouped frequency distribution in Table 2.4). In the interest of variety, let us choose an interval size of 5 for the stem-and-leaf display, and order the stems (intervals) from small to large going down the page. (See Table 2.5.) Each leaf is made up of the scores within a given interval, with each score represented solely by its units digit, and the scores in each leaf are ordered from low to high. Because each observation takes up one space, the stem-and-leaf display provides the same graphic representation as the histogram (rotated 90° counterclockwise), while at the same

TABLE 2.5

Stem-and-leaf display for data in Table 2.3

Stems (Intervals)	Leaves (Observations)	Frequency (f)	Cumulative Frequency (cf)
5–9	9	1	1
10–14	0 1 1 1 3 4 4	7	8
15–19	5 5 5 8 8	5	13
20–24	0 0 1 1 1 1 2 2 2 3 4	11	24
25–29	5 5 6 6 6 6 6 6 8 8 8 8 9 9 9 9	17	41
30–34	0 0 0 0 0 1 1 1 2 2 2 2 2 2 3 3 3 3 4 4	20	61
35–39	5 5 5 7 7 7 7 8 8 8 9 9 9	13	74
40–44	0 1 1 2 3 3 4	7	81
45–49	5 6 7 9	4	85

---

* Note that age, when taken as of one's last birthday, would be an exception to this rule. For example, the midpoint for 10-year-olds would be $10\frac{1}{2}$ years.

† See J. W. Tukey, *Exploratory data analysis* (Reading, Mass.: Addison-Wesley, 1977).

time explicitly giving each score value. Thus you can see both the overall shape of the distribution and such particulars as the largest and smallest observations. The stem-and-leaf display may be supplemented by the frequency distribution and/or the cumulative frequency distribution, as needed.

Since the stem-and-leaf display is used primarily to help comprehend a set of data, you may use or invent whatever variations best accomplish this purpose. For data that cover a wide range of values, you can use the last two digits to represent each score in the leaf (e.g., 460–479 | 63, 68, 74, 74, 78). Or, if there are a great many observations, you might choose to let each leaf entry represent not one but two or more cases.

Frequency Polygons

Regular frequency polygons.   The frequency data in Table 2.4 are presented as a *regular frequency polygon* in Figure 2.2. The strategy is the same as in the case of the histogram: the frequency of a score is expressed by the height of the graph above the X axis, and frequencies are entered on the Y axis and scores on the X axis. The difference is that points, rather than bars, are used for each entry. Thus, the frequency of 5 for the class interval of 9–11 is shown by a dot 5 units up on the Y axis above the score of 10 (the midpoint of the 9–11 interval; with a regular frequency distribution, each score would

FIGURE 2.2

Regular frequency polygon for data in Table 2.4

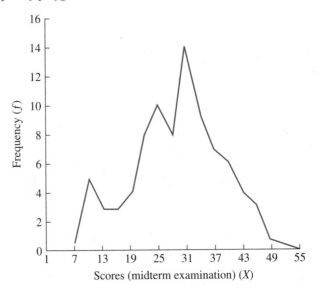

be entered on the *X* axis). All dots are connected with straight lines, and the resulting frequency polygon provides a pictorial illustration of the frequency distribution.

Regular frequency polygons are particularly appropriate for *continuous* data, where results between the score values shown can occur, or could if it were possible to measure with sufficient refinement. Scores, whether or not fractions can occur, are always treated as continuous.

Cumulative frequency polygons. Cumulative frequency distributions are also commonly graphed in the form of frequency polygons, and the resulting figure is called (not very surprisingly) a *cumulative frequency polygon*. An example, based on the data in Table 2.2, is provided in Figure 2.3. Note that, reading from left to right, the cumulative frequency distribution always remains level or increases and can never drop down toward the *X* axis. This is because the cumulative frequencies are formed by successive additions. Thus the *cf* for an interval can be at most equal to, but never less than, the *cf* for the preceding interval.

FIGURE 2.3

Cumulative frequency polygon for data in Table 2.2

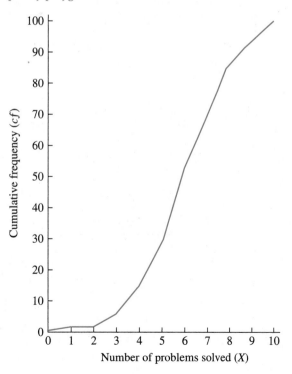

Shapes of Frequency Distributions

It is often useful to talk about the general shape of a frequency distribution. Some important definitions commonly used in this regard are as follows:

Symmetry versus skewness.   A distribution is *symmetric* if and only if it can be divided into two halves, each the "mirror image" of the other. For example, Distributions A, B, C, and G in Figure 2.4 are symmetric. Distributions D, E, F, and H are *not* symmetric, however, since they cannot be divided into two similar parts.

A markedly asymmetric distribution with a pronounced "tail" is described as *skewed* in the direction of the tail. Distribution E is *skewed to the left* (or negatively skewed), since the long tail is to the left of the distribution. Such a distribution indicates that most people obtained high scores but some (indicated by the tail) received quite low scores, as might happen if an instructor gave an exam which proved easy for all but the poorest students.

FIGURE 2.4

Shapes of frequency distributions. (A) Normal curve (symmetric, unimodal).
(B) Symmetric, unimodal. (C) Symmetric, bimodal. (D) Asymmetric, bimodal.
(E) Unimodal, skewed to the left. (F) Unimodal, skewed to the right.
(G) Rectangular. (H) J-curve.

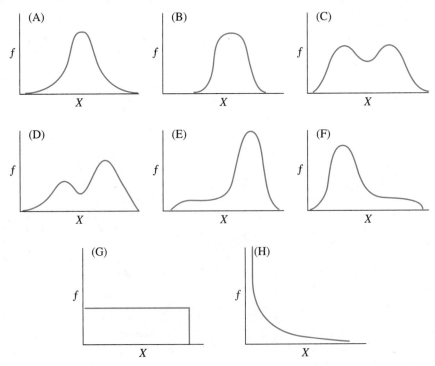

Distribution F, on the other hand, is *skewed to the right* (or positively skewed), as is indicated by the position of the tail. This distribution indicates many low scores and some quite high scores, as could occur if an instructor gave a difficult examination on which only the best students performed respectably.

Modality.   Modality refers to the number of clearly distinguishable high points or "peaks" in a distribution. In Figure 2.4, Distributions A, B, E, and F are described as *unimodal* because there is one peak; Distributions C and D are *bimodal* because there are two clearly pronounced peaks; and Distribution G has no modes at all. Be careful, however, not to be influenced by minor fluctuations when deciding whether to consider a distribution unimodal or bimodal. For example, the curve shown in Figure 2.2 is properly considered as unimodal: with just one less case in the 24–26 interval and one more case in the 27–29 interval, the second "peak" would disappear entirely.

Special distributions.   Certain frequency distributions have names of their own. A particular bell-shaped, symmetric, and unimodal distribution shown in Example A of Figure 2.4 is called the *normal distribution*, about which much more will be said in subsequent chapters. In this distribution, frequencies are greatest at the center, and more extreme scores (both lower and higher) are less frequent. Many variables of interest to the behavioral scientist, such as intelligence, are approximately normally distributed.

Example G illustrates a *rectangular* distribution, in which each score occurs with the same frequency. A distribution approximately like this would occur if one fair die were rolled a large number of times and the number of 1s, 2s, and so on were recorded. (This would, however, be a discrete distribution.)

Example H is an illustration of a *J-curve*, wherein the lowest score is most frequent and the frequencies decrease as the scores become larger. An example of such a distribution would be the number of industrial accidents per worker over a two-month period: the majority of workers would have no accidents at all, some would have one accident, a few would have two accidents, and very few unfortunate souls would have a larger number of accidents. The mirror image of this distribution, which is shaped almost exactly like a "J," is also quite naturally called a J-curve.

Summary

In *regular frequency distributions,* all score values are listed in the first column and the frequency corresponding to each score is listed to the right of the score in the second column. *Grouped frequency distributions* sacrifice some information for convenience by combining several score values in a single *class interval* so that fewer intervals and corresponding frequencies need be listed. *Cumulative frequency distributions* are obtained by starting with the frequency corresponding to the lowest score and adding up the frequencies as you go along. *Histograms* and *frequency polygons* are two common methods for graphing frequency distributions; both can be used with any kind of data, but histograms are particularly appropriate for discrete data and frequency polygons are particularly appropriate for continuous data. The *stem-and-leaf display* combines features of the frequency distribution and histogram and facilitates the comprehension of a data set by seeing it all at once and with as much detail as is needed.

1. Regular frequency distributions

List every score value in the first column, with the highest score at the top. List the *frequency* (symbolized by *f*) of each score to the right of the score in the second column.

2. Grouped frequency distributions

List the *class intervals* in the first column and the frequencies in the second column. It is usually desirable to:

1. Have a total of from 8 to 15 class intervals.
2. Use an interval size of 2, 3, 5, or a multiple of 5, selecting the smallest size that will satisfy the first rule. (All intervals should be the same size.)
3. Make the lowest score in each interval a multiple of the interval size.

Do not use grouped frequency distributions if all scores can be quickly and conveniently reported, because grouped frequency distributions lose information.

3. Cumulative frequency distributions

To the right of the frequency column, form a column of cumulative frequencies (symbolized by *cf*) by starting with the frequency for the lowest score and adding up the frequencies as you go along.

4. Graphic representations

Histograms, in which the frequency of any score is expressed by the height of the bar above that score, are particularly appropriate for discrete data (where results between the score values shown cannot occur).

Stem-and-leaf displays are formed by listing class intervals ("stems") in a column at the left, with specific values within each interval ("leaves") on a horizontal line next to that interval (often represented solely by the units digit).

Frequency polygons, in which the frequency of any score is expressed by the height of the point above that score and points are connected by straight lines, are particularly appropriate for continuous data (where results between the score values shown can occur, or could if it were possible to measure with sufficient refinement).

Chapter 3
Transformed Scores I: Percentiles

PREVIEW

Interpreting a Raw Score

What is a raw score?

Why is it often necessary to compare a raw score to the specific group of scores in which it appears?

What is a transformed score?

Definition of Percentile and Percentile Rank

What is a percentile rank?

What is a percentile?

What is the primary purpose of percentile ranks and percentiles?

Why must we take careful note of the reference group to interpret a percentile rank correctly?

Computational Procedures

How is a raw score transformed into a percentile rank?

Given a percentile, how do we determine the corresponding raw score?

How can percentile ranks be determined more easily when a stem-and-leaf display has been prepared?

Deciles, Quartiles, and the Median

What is a decile?

What is a quartile?

What is the median?

Summary

Interpreting a Raw Score

If you obtain a score of 41 on a 50-point examination, you will need additional information in order to determine how well you did. You can draw some useful conclusions from the fact that your score represents 82% of the total (for example, it is unlikely that you have failed the examination). But you also need to know how your score compares to the specific group of scores in which it appears, namely the scores of the other students in the class. If the examination has proved easy for most students and there are many high scores, your score of 41 may represent only average (or even below average) performance. If the examination was a difficult one for most students, your score may be among the highest (or even the highest).

One way of providing this additional information is to transform the original score (called the *raw score*) into a new score that will show at a glance how well you did in comparison to other students in the class. There are several different kinds of transformed scores. We will discuss one of them — percentiles — in this chapter, and defer a discussion of others (which depend on material in the following chapters) until Chapter 6.

Definition of Percentile and Percentile Rank

A *percentile rank* of a score is a single number that gives the *percent of cases in the specific reference group scoring at or below* that score. If your raw score of 41 corresponds to a percentile rank of 85, this means that 85 percent of your class obtained equal or lower scores than you did, while 15% of the class received higher scores. If instead your raw score of 41 corresponds to a percentile rank of 55, this would signify that your score was slightly above average; 55% of the class received equal or lower scores, while 45% obtained higher scores.

A *percentile* is the score at or below which a given percent of the cases lie. A score that would place you at the 5th percentile would be a cause for concern, since 95% of the class did better and only 5% did as or more poorly.

As these examples illustrate, percentile ranks and percentiles show directly how an individual score compares to the scores of a specific group. In order to interpret a percentile rank correctly, however, you must take careful note of the reference group in question. A college senior who obtains a test score with a percentile rank of 90 would seem to have done well, since his score places him just within the top 10% of some reference group. But if this group consists of high school seniors, the student should not feel proud of his performance! A score at the 12th percentile is usually poor, since only 12% of the reference group did as badly or worse. But if the score was obtained by a high-school freshman and the reference group consists of college graduates, the score may actually represent good performance relative to other high-school freshmen.

It is unlikely that anyone would err in extreme situations such as the foregoing, but there are many practical situations where misleading conclusions can easily be drawn. Scoring at the 85th percentile on the Graduate Record Examination, where the reference group consists of college graduates, is superior to scoring at the 85th percentile on a test of general ability, where the reference group consists of the whole population (including those people not intellectually capable of obtaining a college degree). Conversely, if you score at the 60th percentile on a midterm examination in statistics and a friend in a different class scores at the 90th percentile on her statistics midterm, she is not necessarily superior. The students in her class might be poorer, which would make it easier for her to obtain a high standing in comparison to her reference group. Remembering that a percentile *compares* a score to a *specific group of scores* will help you to avoid pitfalls such as these.

Computational Procedures

Case 1. *Given a raw score, compute the corresponding percentile rank.*

In order to illustrate the computation of percentile ranks, let us suppose that you have received a score of 41 in the hypothetical examination data illustrated in Table 2.3. The grouped and cumulative frequency distributions for these data (Table 2.4) are reproduced in Table 3.1.

To find the percentile rank corresponding to the raw score of 41, do the following:

1. Locate the class interval in which the raw score falls. (This interval has been boxed in Table 3.1.) Let us call this the "critical interval."

TABLE 3.1

Hypothetical midterm examination scores for 85 students: transforming a raw score of 41 to a percentile rank

Class Interval	Frequency (*f*)		Cumulative Frequency (*cf*)
48–50	1		85
45–47	3	8	84
42–44	4		81
39–41	6		77
36–38	7		71
33–35	9		64
30–32	14		55
27–29	8		41
24–26	10	71	33
21–23	8		23
18–20	4		15
15–17	3		11
12–14	3		8
9–11	5		5

2. Combine the frequencies (f) into three categories: those corresponding to all scores *higher* than the critical interval, those corresponding to all scores in the critical interval, and those corresponding to all scores *lower* than the critical interval, as follows:

   As is shown in Table 3.1, a total of 8 people obtained scores higher than the critical interval; 6 people obtained scores in the critical interval; and 71 people obtained scores lower than the critical interval. The last figure is readily obtained by referring to the *cumulative* frequency for the interval just below the critical interval, which shows that 71 people obtained scores of 36–38 or less. Each frequency is then converted to a percent by dividing by N, the total number of people (in this example, 85). We will denote the percent of people scoring in intervals higher than the critical interval by $H\%$ (for *higher*), the percent of people scoring in the critical interval by $I\%$ (for *in*), and the percent of people scoring lower than the critical interval by $L\%$ (for *lower*).

	f	Percent ($=f/N$)
All *higher* intervals	8	8/85 = 9.4% ($H\%$)
Critical interval (39–41)	6	6/85 = 7.1% ($I\%$)
All *lower* intervals	71	71/85 = _83.5%_ ($L\%$)
		Check: = 100.0%

3. It is now apparent that your score of 41 is better than at least 83.5% of the scores, namely those below the critical interval. Thus, your rank in the class expressed as a percentile (or, more simply, your *percentile rank*) must be at least 83.5%. It is also apparent that 9.4% of the scores, the ones above the critical interval, are better than yours. But what of the 7.1% of the scores within the critical interval? It would be too optimistic to assume that your score is higher than the scores of all the other people in the critical interval, since some of these people may also have obtained scores of 41. On the other hand, it would be too pessimistic to assume that you did not do better than anyone in your class interval. The solution is to look at your score in comparison to the size of the interval; the higher your score in relation to the critical interval, the more people in that interval you may assume that you outscored.

In order to determine accurately your standing in the critical interval, you must first ascertain the *lower real limit* of the interval. It may seem as though the lower limit of the 39–41 interval is 39, but appearances are often deceiving. If a score of 38.7 were obtained, in which interval would it be placed? Since this score is closer to 39 than 38, it would be tallied in the 39–41 interval. If a score of 38.4 were obtained, it would be tallied in the 36–38 interval because 38.4 is closer to 38 than to 39. The *real* dividing line between the critical interval of 39–41 and the next lower interval of 36–38 is not 39, but 38.5. Any score between 38.5 and 39.0 belongs in the 39–41 interval, and any score between 38.0 and 38.5 belongs in the 36–38 interval. (For a score of exactly 38.5, it would be necessary to flip a coin or use some other random procedure.) A convenient rule is that the lower real limit of an interval is halfway between the lowest score in that interval (39) and the highest score in the next lower interval (38).

Your score of 41 is 2.5 points (41 − 38.5) up from the lower real limit of the interval. Since the size of the interval is 3, this distance expressed as a fraction is equal to 2.5 points/3 points, or .83 of the interval. Consequently, in addition to the 83.5% of the scores that are clearly below yours, you should credit yourself with .83 of the 7.1% of people in your interval, and your percentile rank is equal to

$$83.5\% + (.83)(7.1\%) = 83.5\% + 5.9\%$$

$$= 89.4\%$$

This procedure is conveniently summarized by the following formula:

$$\textit{percentile rank} = L\% + \left(\frac{Score - LRL}{h} \cdot I\% \right)$$

where

L% and I% are obtained from step 2 (page 35)
Score = raw score in question
LRL = lower real limit of critical interval
h = interval size

In our example, this is equal to

$$83.5\% + \left(\frac{41 - 38.5}{3} \cdot 7.1\% \right) = 83.5\% + 5.9\%$$

$$= 89.4\%$$

So your percentile rank is equal to 89.4%, indicating that approximately 89% of the class received equal or lower scores and only about 11% received higher scores. This result is depicted in Figure 3.1.

This procedure is also suitable for use with regular frequency distributions, where the interval size (h) equals 1. You will find that the fraction

$$\left(\frac{Score - LRL}{h} \right)$$

always equals one-half for regular frequency distributions.

When you are able to use a regular frequency distribution, you will get somewhat more accurate results, since grouped frequency distributions lose information. For purposes of descriptive statistics, however, the loss of

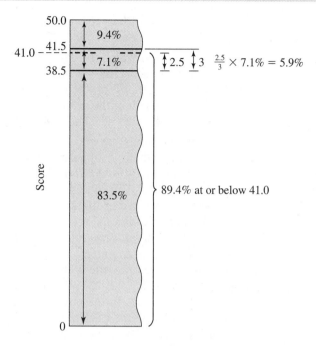

Illustration of percentile rank corresponding to a raw score of 41 for data in Table 3.1

Note: This percentile rank is computed for a particular *point* on the scale—that is, a score of exactly 41.00. The fact that a score of 41 actually occupies the interval 40.5 to 41.5 is deliberately disregarded in this instance so that we can talk of a percentage above and a percentage at or below the given point, which together sum to 100%. Thus, the point is a "razor-thin" dividing line which divides the group into two parts, and not an interval in which scores may fall.

accuracy incurred by grouping will usually not be important, and you can follow the guidelines for forming frequency distributions given in the preceding chapter.

Case 2.   Given a percentile, compute the corresponding raw score.

In the previous problem, you had a raw score (41) and wished to find the corresponding percentile rank. It is also useful to know how to apply the percentile procedures in reverse—to find the raw score that corresponds to a specified percentile value. For example, suppose that an instructor wishes to recommend remedial procedures to the bottom 25% of the class. What raw score should be used as the cutting line? In this example, the percentile (25%) is specified and the raw score is needed, and the steps are as follows:

1. Convert the percentile to a case number by multiplying the percentile by *N*. In the present example, this is equal to (.25) × (85) or 21.25. Thus, the score that corresponds to the individual whose rank is 21.25 from the

bottom of the class (the person scoring at the 25th percentile) is the cutting line that you need.

2. Find the interval in which the case number computed in step 1 falls. This is easily accomplished by starting at the *bottom* of the *cumulative* frequency distribution and proceeding upward until you find the *first* value equal to or greater than the critical case (21.25); the corresponding interval is the "critical interval." This step is illustrated in Table 3.2.

3. The 21.25th case must have a score of at least 20.5, the lower real limit of the interval in which it appears. However, the critical interval covers three score points (from 20.5 to 23.5); what point value corresponds to the 25th percentile? The solution lies in considering how far up from the lower end of the interval the 21.25th case falls, and assigning an appropriate number of additional score points. If the 21.25th case falls near the bottom of the interval, very little will be added to 20.5; if the 21.25th case falls near the top of the interval, a larger quantity will be added to 20.5. In our present example, there are 15 cases *below* the critical interval, so the 21.25th case falls $(21.25 - 15)$ or 6.25 cases up in the interval. The total number of cases in the interval is 8, so this distance expressed as a fraction is 6.25 cases/8 cases or .78. Therefore, in addition to the lower real limit of 20.5, .78 of the three points included in the critical interval must be added in order to determine the point corresponding to the 21.25th case. The desired cutting score is therefore equal to

$$20.5 + (.78 \times 3) = 20.5 + 2.3$$

$$= 22.8$$

TABLE 3.2

Hypothetical midterm examination scores for 85 students: finding the raw score corresponding to the 25th percentile

Class Interval	Frequency (f)	Cumulative Frequency (cf)	
48–50	1	85	
45–47	3	84	
42–44	4	81	
39–41	6	77	
36–38	7	71	
33–35	9	64	
30–32	14	55	
27–29	8	41	
24–26	10	33	
21–23	8 = f	23	first $cf \geq 21.25$
18–20	4	15	
15–17	3	11	
12–14	3	8	
9–11	5	5	

15 = SFB (bracketing 4, 3, 3, 5)

$pN = .25 \times 85 = 21.25$

This procedure is conveniently summarized by the following formula:

$$Score_p = LRL + \left(\frac{pN - SFB}{f} \cdot h \right)$$

where

$$Score_p = \text{score corresponding to the } p\text{th percentile}$$
$$LRL = \text{lower real limit of critical interval}$$
$$p = \text{specified percentile}$$
$$N = \text{total number of cases}$$
$$SFB = \text{sum of frequencies } below \text{ critical interval}$$
$$f = \text{frequency within critical interval}$$
$$h = \text{interval size}$$

In our example,

$$Score_{.25} = 20.5 + \left(\frac{(.25)(85) - 15}{8} \cdot 3 \right)$$

$$= 20.5 + \left(\frac{6.25}{8} \cdot 3 \right)$$

$$= 22.8$$

Students with scores of 22 or less are assigned to remedial work, and those with scores of 23 or more are not. The calculations are depicted in Figure 3.2.

Avoid the temptation to use in the formula any percent that happens to be handy. For example, suppose that the top 10% of the class is to receive a grade of A. What cutting line should be used? Before doing any calculations, you must note that the "top 10%" corresponds to the *90th percentile*; therefore, $p = .90$. The solution:

$$pN = 76.5$$

$$\text{Critical interval} = 39-41$$

$$Score_{.90} = 38.5 + \frac{76.5 - 71}{6} \cdot 3$$

$$= 38.5 + 2.75$$

$$= 41.2$$

This result should not prove surprising, since we found in the previous section that a score of 41.00 corresponded to the 89.4th percentile. You will

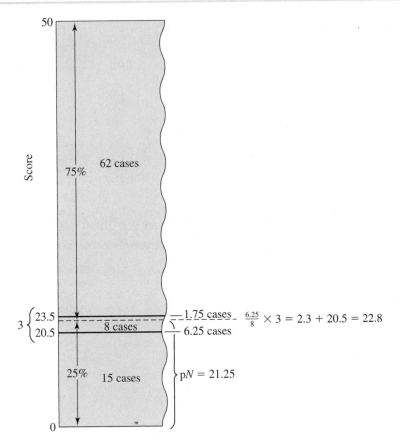

Illustration of raw score corresponding to percentile rank of 25% for data in Table 3.2

obtain a surprising (and erroneous) result, however, if you unthinkingly set p equal to .10.

Alternative Method Using Stem-and-Leaf Displays

An easier procedure for determining percentile ranks (Case 1) is available when a stem-and-leaf display has been prepared, since it offers a ranking of all the observations. Given Table 2.5, which provides a stem-and-leaf display of the examination scores, all that is necessary to ascertain a percentile rank is some counting.

For example, the raw score of 41 appears in the second stem from the bottom (40–44). It is exceeded by four scores in that stem and four more scores in the last stem (45–49), a total of eight scores. Since there are 85 scores in all, there are $85 - 8 = 77$ scores at or below a score of 41. Therefore, the raw score of 41 has a corresponding percentile rank of

$77/85 = 90.6\%$. Why does this not agree with the value of 89.4% obtained from the formula in the preceding section? Because the latter interpolates within the interval, treating its scores as though they were uniformly distributed in that interval. Conversely, with the stem-and-leaf display, you determine the percentile rank exactly. This discrepancy is usually not important, however, so you might well decide not to expend the effort to make up a stem-and-leaf display solely for the purpose of computing a percentile rank. The primary advantage of this method occurs when such a display already exists (or is needed for other purposes), and when the number of observations is small enough to make such a display feasible (say, less than 100).

If a stem-and-leaf display has been prepared, the determination of the raw score corresponding to a given percentile (Case 2) is also simplified. Thus, for $p = .25$, first compute $pN = .25(85) = 21.25$ (as before). Now refer to the cumulative frequency distribution included with the stem-and-leaf display in Table 2.5, and observe that the 21.25th score occurs in the $20-24$ stem (interval). Since 13 scores fall below this stem, you need to cumulate an additional 8.25 scores $(21.25 - 13)$. Therefore, simply count along the leaf to the 8.25th score. The eighth score is the second 2 (standing for 22) and the ninth score also happens to be 2, so the 8.25th score—and, therefore, the score corresponding to the 25th percentile—must also be 22.

Deciles, Quartiles, and the Median

Certain percentile values have specific names, as follows:

Percentile		Decile		Quartile		
90th	=	9th				
80th	=	8th				
75th	=			3rd		
70th	=	7th				
60th	=	6th				
50th	=	5th	=	2nd	=	MEDIAN
40th	=	4th				
30th	=	3rd				
25th	=			1st		
20th	=	2nd				
10th	=	1st				

Whereas the percentile divides the total number of cases into 100 equal parts, the *decile* divides the number of cases into 10 equal parts and the *quartile* divides the number of cases into four equal parts. The score corresponding to the 50th percentile has the unique property that exactly half the scores in the group are higher and exactly half the scores in the group are equal or lower. It is called the *median,* and is one of the measures of central tendency that will be discussed in the next chapter.

Summary

The *percentile rank* corresponding to a given score refers to the percent of cases in a given reference group *at or below* that score. A specified raw score may be converted to the corresponding percentile rank to express its standing relative to the reference group, or the raw score corresponding to a specified percentile may be determined, as follows:

1.   *Case 1. Given a score, find the corresponding percentile rank (PR).*

   1. Find the class interval in which the score falls.
   2. Set up the following diagram and fill in the missing values:

	f	Percent ($=f/N$)
All *higher* intervals	___	$H\% = $ ___
Interval in which score falls	___	$I\% = $ ___
All *lower* intervals	___	$L\% = $ ___
		Check $= 100\%$

   3. Let

   LRL = *lower real limit* of the interval in which the score falls
   (The lower real limit of an interval is the number halfway between the lowest number in that interval and the highest number in the next lower interval.)
   h = interval size

   Then,

$$PR = L\% + \left(\frac{Score - LRL}{h} \cdot I\% \right)$$

2.   *Case 2. Given a percentile (p), find the corresponding raw score (Score$_p$).*

   1. Compute $p \times N =$ Nth case.
   2. Find the interval in which this case falls.
   3. Let LRL = lower real limit of this interval
       SFB = *sum of frequencies below* this interval
       f = frequency within this interval
       h = interval size

   Then,

$$Score_p = LRL + \frac{pN - SFB}{f} \cdot h$$

Chapter 4
Measures of Central Tendency

PREVIEW

Introduction

What is a measure of central tendency?

What do we gain by using a measure of central tendency instead of a regular frequency distribution?

What important information is *not* conveyed by a measure of central tendency?

The Mean

How do we compute the mean of a set of scores?

Is the population mean computed any differently from the sample mean?

How is the mean computed from a regular frequency distribution?

What must the sum of the deviations of all scores from the mean be equal to? What does this imply about the way in which very large or very small scores affect the mean?

When should the mean be used as the measure of central tendency?

The Median

What is the median and how is it computed?

How do very large or very small scores affect the median?

In what kind of distribution is the median larger than the mean?

In what kind of distribution is the mean larger than the median?

When should the median be used as the measure of central tendency?

The Mode

What is the mode?

Why is the mode usually *not* used as the measure of central tendency?

Summary

Introduction

The techniques presented in Chapter 2 are useful when you wish to provide a detailed summary of all of the data in a convenient format. Often, however, your primary objective will be to highlight certain important characteristics of a group of data. For example, suppose that an inquisitive friend wants to know how well you are doing in college. You might hastily collect all of your semester grade reports, and compile a regular frequency distribution such as the following:

Grade	f
A	4
B	9
C	6
D	1
F	0

This would show that you received four grades of A, nine grades of B, and so forth. But all this detail is not essential in order to answer the question, and it would undoubtedly prove tiresome both for you and for your audience. In addition, presenting the data in this form would make it awkward for your friend to compare your performance to his own college grades. A better plan would be to select one or two important attributes of this set of data, and summarize them so that they could be reported quickly and conveniently.

One item of information that you would want to convey is the general *location* of the distribution of grades. You could simply state that your college work was slightly below the B level. If you wished to be precise, you would report your numerical grade point average—a single number that describes the general location of this set of scores. In either case, you would sum up your performance by referring to a central point of the distribution. It would be misleading to describe your overall performance as being at the A or D level, even though you did receive some such grades.

This is one of many situations that benefit from the use of a *measure of "central tendency"*—*a single number that describes the general location of a set of scores.* Other examples include the average income of families in the United States, the number of cents gained or lost by an average share of stock on the New York Stock Exchange in a single day, and the number of seconds taken by the average rat to run a T maze after 24 hours of food deprivation.

It should be stressed, however, that *the location of a set of data is not its only important attribute.* Suppose that the average score on a statistics quiz that you have just taken is 5.0. This average provides information as to the general location of the set of quiz scores, but it does not tell you how many high and low scores were obtained. Consequently, you cannot determine what score will be needed to ensure an A (or to just pass with a D!). In a distribution such as the following one, a score of 7 would rank very highly:

$$7\ 7\ 6\ 5\ 4\ 4\ 4\ 3$$

On the other hand, a score of 7 would not seem so illustrious in a distribution like this:

$$10\ 10\ 9\ 7\ 5\ 4\ 3\ 2\ 0\ 0$$

In both examples, however, the average, measured as the mean, is equal to 5.0. This indicates that there are important aspects of a set of data which are *not* conveyed by a measure of central tendency; a second vital characteristic will be considered in the next chapter.

The Mean

Computation. The *mean* of a set of scores is computed by adding up all the scores and dividing the result by the number of scores. In symbols,

$$\overline{X} = \frac{\sum X}{N}$$

where

$$\overline{X} = \textbf{sample mean}$$
$$\sum X = \textbf{sum of the } X \textbf{ scores (see Chapter 1)}$$
$$N = \textbf{total number of scores}$$

In the case of the first set of quiz scores given above, the mean is equal to 5.0, as shown:

$$\frac{7 + 7 + 6 + 5 + 4 + 4 + 4 + 3}{8} = \frac{40}{8} = 5.0$$

Note that simply computing $\sum X$ is not sufficient to identify the location of these scores. What is further required is to divide by N (the number of scores). This step ensures that the means of two different samples will be comparable, even if they are based on different numbers of scores. The mean of the second set of quiz scores presented in the previous section is equal to 50/10 or 5.0; $\sum X$ is different, but the mean correctly shows that the location is the same.

In the case of populations, the Greek letter mu, μ, is used to represent the *population mean.* Nevertheless, the procedure is the same—sum all scores and divide by N.*

Computation from a regular frequency distribution.   If scores are available in the form of a regular frequency distribution, the mean is most easily computed from the following formula:

$$\overline{X} = \frac{\sum fX}{N}$$

where

$\overline{X}$ = sample mean
fX = X score multiplied by the frequency of that score
$\sum fX$ = sum of fX values
N = total number of scores

Perhaps because of the extra symbol f in the numerator, this equation is often a source of confusion. It gives exactly the same result as would the preceding formula applied to the same data in an untabulated format. It may be helpful to think of the first formula for the sample mean as a special case of the second in which $f = 1$.

The two procedures for computing the mean are compared in Figure 4.1. Note that the value $N = 20$ is easily recovered from the regular frequency distribution by computing $\sum f$: the total of the *frequencies* shows *how many scores* there are.

As was mentioned in Chapter 2, means (and other statistics) should in general not be computed from *grouped* frequency distributions, which do not give the exact value of every score. They may be approximated by treating all the scores in any interval as if they fell at the midpoint of the interval.

Interpretation.   One important characteristic of the mean is that the sum of distances (or *deviations*) of all scores from the mean is zero. That is,

$$\sum\left(X - \overline{X}\right) = 0$$

It can readily be proved that this must always be true.† As an illustration, consider once again the small set of quiz scores discussed previously.

---

* The distinction between samples and populations was introduced in Chapter 1, and it will be treated at greater length in Chapter 8.

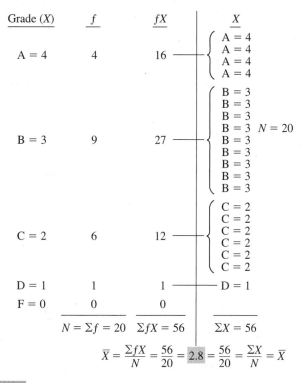

Grade (X)	f	fX	X
A = 4	4	16	A = 4 A = 4 A = 4 A = 4
B = 3	9	27	B = 3 B = 3 B = 3 B = 3 B = 3 $N = 20$ B = 3 B = 3 B = 3 B = 3
C = 2	6	12	C = 2 C = 2 C = 2 C = 2 C = 2 C = 2
D = 1	1	1	D = 1
F = 0	0	0	
	$N = \Sigma f = 20$	$\Sigma fX = 56$	$\Sigma X = 56$

$$\overline{X} = \frac{\Sigma fX}{N} = \frac{56}{20} = \boxed{2.8} = \frac{56}{20} = \frac{\Sigma X}{N} = \overline{X}$$

FIGURE 4.1

The mean computed from a regular frequency distribution and from untabulated data

Score	Deviation from Mean $(X - \overline{X})$	
7	+2	
7	+2	
6	+1	
5	0	$(\overline{X} = 5)$
4	−1	
4	−1	
4	−1	
3	−2	
	$\Sigma (X - \overline{X}) = +5 - 5 = 0$	

The mean balances or equates the sums of the positive and negative deviations. It is in this sense that it gives the location of the distribution. This

---

† $\Sigma (X - \overline{X}) = \Sigma X - N\overline{X}$ (Rule 7, Chapter 1)

but $\Sigma X = N\overline{X}$

because $\overline{X} = (\Sigma X)/N$. Therefore,

$\Sigma (X - \overline{X}) = N\overline{X} - N\overline{X}$

     $= 0$

implies that the mean will be sensitive to extreme values on one side that are not balanced by extreme values on the other side, as the following example shows:

Score	$(X - \bar{X})$	
39	+ 30	
7	− 2	
6	− 3	
5	− 4	$(\bar{X} = 9)$
4	− 5	
4	− 5	
4	− 5	
3	− 6	
$\Sigma (X - \bar{X}) = +30 - 30 = 0$		

As a result of changing one score from 7 to 39, the mean shows a substantial four-point increase. In addition, all of the other seven scores now fall below the mean in order to balance the effects of the large positive deviation introduced by the score of 39. One might well question the use of the mean to describe the location of a set of data in a situation where it is so influenced by one extreme score!

As another example of the sensitivity of the mean to unbalanced extreme values, consider the case of an unethical manufacturing company in which the president earns one million dollars per year and the 99 assembly line workers earn only $6,500 per year. The president might attempt to refute criticism of the company's miserly tactics by arguing that the mean annual income of these 100 people is $16,435. The workers would undoubtedly object to the appropriateness of this figure, which is more than $2\frac{1}{2}$ times their actual salaries! Here again, the use of the mean as the index of location is questionable. It almost always is with income data, since almost all income distributions are positively skewed.

Usage. The mean has many advantageous properties. It takes all of the scores into account, so it makes the most of the information provided by the data. Also, the mean is the most stable of the measures of central tendency for most distributions encountered in practice: it is the most consistent across different samples drawn from the same population. For this and other reasons, many of the procedures of inferential statistics make use of the mean. Thus, a third advantage of the mean is that it is usable as a datum in further statistical analyses, while other measures of central tendency usually are not. For these reasons, the mean is the most frequently suitable measure of central tendency.

At times, however, the first advantage becomes a liability instead of an asset. As we have seen, extreme scores at one end of a distribution exert a strong influence on the mean and cause it to give a misleading picture of the location of the distribution. Therefore, when a distribution is highly skewed and when you do not intend to use the measure of central tendency in subsequent statistical analyses, you should seek an alternative to the mean that will not be affected by unbalanced extreme scores.

In some cases, the actual size of the extreme scores may be unknown. For example, suppose that a few subjects in a learning experiment do not learn the task even after a great many trials. You might have to terminate the experiment for anyone failing to learn after (say) 75 trials. Such subjects would therefore have learning scores of "at least 75 trials." They would be the slowest learners (and should not be discarded from the experiment, as the results would then be biased); but their exact scores would be unknown, so you could not compute a sample mean.

In situations such as these, a measure of central tendency is needed that does *not* take the exact value of extreme scores into account. Such a measure is available, and it is called the median.

The Median

Computation. The median (Mdn) is defined as the score corresponding to the 50th percentile. It is computed using the procedures given in the previous chapter (simply compute $Score_{.50}$).

As an illustration, the computation of the median for the "grade" data given at the beginning of this chapter is shown in Figure 4.2. There are 20 observations in all, so the median is the score such that 10 cases (*half* of the total) fall above it and 10 cases fall below it. If you were to take the score of

FIGURE 4.2

The median computed from a regular frequency distribution

X	f	cf		
A = 4	4	20		
B = 3	9	16	first $cf \geq 10$	
C = 2	6	7		
D = 1	1	1		
F = 0	0	0		

$p = .50, N = 20$
$pN = (.50)(20) = 10$
"Critical interval" (c.i.)
is the "B" interval.

$LRL = 3.0 - 0.5 = 2.5$
$SFB = 7$ (cases below c.i.)
$f = 9$ (cases in c.i.)
$h = 1$ (interval size)

$$Mdn = Score_{.50}$$
$$= LRL + \frac{pN - SFB}{f} \cdot h$$
$$= 2.5 + \frac{(.50)(20) - 7}{9} \cdot 1$$
$$= 2.5 + \frac{3}{9}$$
$$= 2.83$$

2.50 (the dividing line between B and C) as your median, there would be only 7 cases below it (6 Cs and 1 D) and 13 cases above it (4 As and 9 Bs). So you must also take 3 of the 9 cases from the B interval and add them to the group below the cutting line, and this is accomplished by moving up 3/9 of a point from the starting place of 2.50. Thus the median is equal to 2.50 + 3/9, or 2.83.

Interpretation. The median is the *middle* score in the distribution when scores are put in order of size. If there is an even number of scores (so that there is no single middle score), the median is computed by averaging the two middle scores (as shown in the examples that follow).

When the distribution of scores is symmetric, the mean and the median will be equal (as is almost exactly true for the "grade" data). In a *positively skewed* distribution, where there are extreme values at the higher end, the mean will be pulled upward by the extreme high scores and will therefore be larger than the median:

$$3\ 4\ 4\ 4\ 5\ 6\ 7\ 39$$

We have seen that the *mean* of these data is 9.0. The *median* is only 4.5, however; it is *not* affected by the size of any extremely large (or extremely small) values. In fact, the median will remain 4.5 even if the value of 39 is changed to a huge number such as 39,000. (The median will also remain 4.5 if the value of 3 is changed to an extremely small number, such as minus one million. But the median *will* change somewhat if the value of 39 is changed to a small number like 3, for there will no longer be 4 cases above it and 4 cases below it.) Similarly, in a learning experiment where some subjects fail to learn the task even after the full 75 trials, you would call the highest class interval "75 trials or more" and compute the median, which is not affected by the numerical value of extremely high (or low) scores.

In a *negatively skewed* distribution where the extreme values are at the lower end, the mean will be pulled downward by the extreme low scores and will therefore be smaller than the median:

$$3\ 6\ 25\ 26\ 27\ 27\ 27\ 29$$

Here the mean is equal to 21.25, while the median is equal to 26.5.

Usage. Use the median when either (1) the data are highly skewed or (2) there are inexact data at the extremes of the distribution. This will enable you to profit from the fact that the median is not affected by the size of extreme values. In almost all other cases, use the mean so as to benefit from its numerous advantages. Except for a few infrequent situations that require the median (Chapter 18), the mean is the measure of location that is usually used in inferential statistics, with the median reserved for situations where the objectives are purely descriptive.

The Mode

The mode is the score that occurs most often. For example, the mode of the data in Figure 4.1 is 3 (or B) since this score was obtained more often than any of the others (9 times).

The mode is a crude descriptive measure of location that ignores a substantial part of the data. Therefore, it is not usually used in research in the behavioral sciences. If you were to record the results of a roulette wheel on a few thousand trials, however, your best bet on subsequent spins would be the mode, in the hope that you had identified the single case most likely to occur in one spin of a flawed wheel. (If the wheel is unbiased, the mode of the previous trials is as good — or as bad — a choice as any other.)

Summary

One important attribute of a set of scores is its *location:* where in the possible range between minus infinity and plus infinity the scores tend to fall. This can be described in a single number by using either the *mean,* the best measure in most instances; or the *median* (the score corresponding to the 50th percentile), preferable when data are highly skewed or there are extreme data whose exact values are unknown, and when the objectives are purely descriptive.

1. The sample mean

$$\bar{X} = \frac{\sum X}{N}$$

1. Use the formula $\bar{X} = (\sum fX)/N$ with regular frequency distributions.
2. The sample mean $(\bar{X})$ is an estimate of the population mean (μ).

2. The median

For either grouped or ungrouped data, compute the score corresponding to the 50th percentile. Recall from Chapter 3 that:

$$Score_{.50} = \text{Mdn} = LRL + \frac{.50N - SFB}{f} \cdot h$$

3. The mode

This is the most frequently obtained score. The mode is at best a rough measure and is generally less useful than the mean or median.

Chapter 5
Measures of Variability

In addition to general location, there is a second important attribute of a distribution of scores—its *variability*. Measures of variability are used extensively in the behavioral sciences, so it is essential to understand the meaning of this concept as well as the calculational procedures.

The Concept of Variability

Variability refers to how *spread out or scattered* the scores in a distribution are (or, how like or unlike each other they are). As an illustration, a few distributions involving a small number of scores are shown below.

Distribution					
1	2	3	4	5	6
7.0	7.2	40.2	7.0	10.0	97.8
7.0	7.1	40.1	7.0	10.0	88.5
7.0	7.1	40.1	6.0	9.0	83.4
7.0	7.1	40.1	5.0	7.0	76.2
7.0	7.1	40.1	4.0	5.0	69.9
7.0	7.0	40.0	4.0	4.0	67.3
			4.0	3.0	58.4
			3.0	2.0	44.7
				0.0	
				0.0	

The minimum possible variability is zero. This will occur only if all of the scores are exactly the same, as in Distribution 1, and there is no variation at all.

In Distribution 2, there is a very small amount of variability. The scores are somewhat spread out, but only to a very slight extent.

Distribution 3 is equal in variability to Distribution 2. The locations of these two distributions differ, but variability is not dependent on location. The distance between each score and any other, and hence the amount of spread, is identical.

Each of the remaining three distributions is more variable than the ones that precede it. At the opposite extreme to Distribution 1 would be a distribution with scores spread out over the entire range from −1,000,000 to +1,000,000 (or more). Such extreme variabilities, however, are rarely encountered in practice.

Variability is important in many areas, although it is frequently not reported (or described vaguely in words) because it is less familiar to nontechnical audiences than is central tendency. We list a few examples:

Testing. Suppose that you score 75 on a statistics midterm examination and that the mean of the class is 65; the maximum possible score is 100. Although your score cannot be poor because it is above average, its worth in comparison to the rest of the class will be strongly influenced by the variability of the distribution of examination scores. If most scores are clustered tightly around the mean of 65, your score of ten points above average will stand out as one of the highest (and may well merit a grade of A). But if the scores are widely scattered, and values in the 80s and 90s (and 40s and 50s) are frequently obtained, being ten points above average will not be exceptional because many will have done better. In this case, your score may be worth no more than a B− or C+. A mean of 65 together with low variability would indicate that you did very well, whereas a mean of 65 together with high variability would imply that your performance was less than outstanding compared to the group that took the test. These two possibilities are illustrated graphically in Figure 5.1. (Note that the spread of a frequency polygon indicates the variability of the distribution.)

Consider Distributions 4 and 5, which were discussed in Chapter 4. The mean of both distributions is 5.0. Yet a score of 7 ranks higher in Distribution 4, where the scores are less variable, than in Distribution 5, where the scores are more variable. Looking at the lower end of these distributions, a score of 3 is poorer in comparison to the group in Distribution 4. Being two points above or below average stands out more in a distribution in which there is less spread.

Sports. Suppose that two professional basketball players average 20.3 points per game. Although their mean performance is the same, they may be different in other respects. One player may be consistent (low variability) and always score close to 20 points in each game, rarely hitting as high as 25 but also rarely falling to 15 or less. The second player may be very erratic (high variability); he scores more than 30 points in some games, but drops to below 10 in others. The second player is likely to be a much greater

FIGURE 5.1

Frequency polygons of two distributions with the same mean but different variability

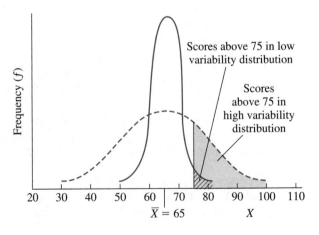

source of frustration to his coach and to the fans. Here again, a measure of variability would provide useful additional information to that given by the mean.

Psychology. Modern psychology is based on the idea that people differ. On any trait of interest — musical ability, throwing a baseball, introversion, height, mathematical ability — people are distributed over the entire range from low to high. If all people were the same, the behavior of the entire population could be predicted from a knowledge of one individual, and psychological research would be unnecessary; anything true about you would be true about everyone else. Since this is not the case, the psychologist has the essential task of measuring and explaining variation — why some people are more neurotic and others less so, why some people do better in school or on the job than others, why one person performs differently on different occasions, and so on. Thus variability is the *raison d'être* of the psychologist.

Statistical inference. Another reason for the importance of variability will become evident when statistical inference is discussed (Chapters 8ff.). Other things being equal, the more variable a phenomenon is, the less precise is the estimate you can get of the population's location (for example, mean) from sample information.

Rather than describe variability in ambiguous terms such as "small" or "large," it is preferable to summarize the variability of a distribution of scores in a single number. Techniques for accomplishing this are discussed next.

The Range

One possible way of summarizing the variability of a distribution is to look at the distance between the smallest and largest scores. The *range* of a distribution is defined as the largest score minus the smallest score. For example, the range of Distribution 4 is (7.0 − 3.0) or 4.0.

While this procedure makes intuitive sense, it is likely to give misleading results on many occasions because the extreme values are frequently atypical of the rest of the distribution. Consider the following two distributions:

DISTRIBUTION A: 10 10 10 9 7 6 5 4 4 3 2 0 0

DISTRIBUTION B: 10   6   6 5 5 5 5 5 5 4 4 0

In both of these examples, the range is equal to 10. Yet Distribution A, with scores spread out over the entire 10-point range, is more variable than Distribution B, where all but two scores are concentrated near the middle of the distribution. The range is a poor measure of the variability of Distribution B because the extreme values are not typical of the total variation in the distribution; if the two extreme scores are excluded, the range drops to 2. Since this type of distortion occurs fairly often, the range (like the mode) is best regarded as a crude measure that should not generally be used in behavioral science research.

The Standard Deviation and Variance

We have seen that the mean, an average that takes all of the scores into account, is usually the best measure of central tendency. Similarly, an "average" variability that is based on all the scores will usually provide the most accurate information. Before an *average* variability can be computed, however, this concept must be defined in terms of an individual score.

Variability actually refers to the difference between each score and every other score, but it would be quite tedious to compute this in practice (especially if N is large). If there are 100 scores, you would have to compute the difference between the first score and each of the 99 other scores, compute the difference between the second score and each of the 98 remaining scores, and so on—4,950 differences in all.

A more feasible plan, which will serve the purpose equally well, is to define the "differentness" or *deviation* of a single score in terms of how far it is from the center of the distribution. In a distribution of scores that is closely packed together, most scores are close to each other and hence close to their center. Conversely, in a highly variable distribution, some scores are quite a distance from each other and hence far from their center. Since the mean is the most frequently used measure of central tendency, a reasonable procedure is to define the deviation of a single score as its *distance from the mean:*[*]

$$\text{Deviation score} = X - \overline{X}$$

Extremely deviant scores (ones far away from the mean) will have numerically large deviation scores, while scores close to the mean will have numerically small deviation scores.

The next step is to derive a measure of variability that will take into account the deviations of all of the scores. There are several possible ways to do this. If we were to average the deviation scores by the usual procedure of summing and dividing by N, we would get

$$\frac{\sum (X - \overline{X})}{N}$$

---

[*] A reference point other than the mean could be used, but the choice of the mean has certain statistical advantages as well as making good intuitive sense. For example, it can be proved that the mean is the value of c about which $\sum (X - c)^2$, the sum of squared deviations, is a minimum (the importance of which will become apparent in the following discussion). The mean, however, is *not* the point about which the sum of deviations whose signs are ignored is a minimum; the *median* is this point.

It will prove extremely frustrating to try and use this as the measure of variability because, as was proved in Chapter 4, $\sum (X - \overline{X})$ is *always* equal to zero. As a result, this "measure" cannot provide any information as to the variability of any distribution.

This problem could be overcome if all of the deviations were positive. We might therefore first take the *absolute value* of each deviation, its numerical value ignoring the sign, and then compute the average variability. That is, we might compute

$$\frac{\sum |X - \overline{X}|}{N}$$

where $|X - \overline{X}|$ is the absolute value of the deviation from the mean. This is not an unreasonable procedure (and in fact yields a descriptive measure called the *average deviation*), but it is usually rejected because absolute values are unsuitable for use in further statistical analyses. The measure that is most frequently used circumvents this difficulty by *squaring* each of the deviations prior to taking the average. The sum of the squared deviations from the mean, $\sum (X - \overline{X})^2$, is symbolized by *SS* and is called the *sum of squares*. The measure of variability produced by taking the average of the sum of squares is called the *variance,* and is symbolized by σ^2:

$$\sigma^2 = \frac{\sum (X - \overline{X})^2}{N} = \frac{SS}{N}$$

This is a basic measure of the variability of any set of data. However, when the data of a sample are used to estimate the variance of the population from which the sample was drawn, the *population variance estimate* (symbolized by s^2) is computed instead:

$$s^2 = \frac{\sum (X - \overline{X})^2}{N - 1} = \frac{SS}{N - 1}$$

In each of the above formulas, the order of operations is: (1) subtract the mean from each score; (2) square each result; (3) sum; (4) divide. The sample estimate of the population variance, s^2, is computed somewhat differently from σ^2; the sum of squared deviations is divided by $N - 1$ instead of N. This is to enable s^2 to be an *unbiased* estimate of the population variance—that is, an estimate that on the average will be too large as often as it is too small.

There is one remaining difficulty. Having squared the deviations to eliminate the negative numbers that otherwise would have led to a total of zero with annoying regularity, the variance is in terms of original units *squared.* For example, if you are measuring IQ, the variance indexes variability in terms of *squared IQ deviations.* It is frequently preferable to have a measure of variability that is in the same units as the original measure, and this can be accomplished by taking the positive square root of the variance. This yields a commonly used measure of variability called the *standard deviation,* symbolized by σ or s depending on whether the variance or the population variance estimate is used:

$$\sigma = +\sqrt{\sigma^2} = \sqrt{\frac{\sum (X - \overline{X})^2}{N}} = \sqrt{\frac{SS}{N}}$$

$$s = +\sqrt{s^2} = \sqrt{\frac{\sum (X - \overline{X})^2}{N - 1}} = \sqrt{\frac{SS}{N - 1}}$$

Whereas the mean represents the "average *score,*" the standard deviation represents a kind of *"average variability"*—the average of the deviations of each score from the mean $(X - \overline{X})$—with two minor complications: squaring and subsequently taking the positive square root, to eliminate the minus signs before averaging and return to the original unit of measurement afterwards; and dividing by $N - 1$ instead of N when estimates of the population are involved. The formulas given above are called the *definition* formulas for σ and s because their primary function is to define the meaning of these terms; they are not necessarily the formulas by which σ and s are most easily computed.

Illustrative examples of the computation of σ^2 and σ, and s^2 and s using the definition formulas are shown on the left-hand side of Table 5.1. If the eight scores in Table 5.1 represent the entire population, σ is computed; while if these eight scores are a sample from a larger population, s is computed. Note that a partial check on the calculations is possible in that $\sum (X - \overline{X})$ should always equal zero. As expected from the previous discussion concerning these distributions, Distribution 5 has a larger standard deviation (is more variable) than Distribution 4. The "average" deviation from the mean is 3.66 points in Distribution 5, but is only 1.5 points in Distribution 4.

Computing formulas. Using the definition formulas to calculate σ and s can be awkward for several reasons. If the mean is not a whole number, subtracting it from each score will yield a deviation score with decimal places, which when squared will produce still more decimal places. This can make the computations quite tedious. Second, the mean must be calculated before the formula can be used, which necessitates two steps in the computation of the standard deviation (although normally the mean will be desired anyway). Therefore, *computing formulas* for σ and s have been derived by manipulating the definition formulas algebraically. The computing formulas and definition formulas yield identical results, but the computing formulas are designed to facilitate calculator computation. The computing formulas for the standard deviation and variance are:

$$\sigma^2 = \frac{1}{N}\left[\sum X^2 - \frac{(\sum X)^2}{N}\right], \qquad \sigma = \sqrt{\frac{1}{N}\left[\sum X^2 - \frac{(\sum X)^2}{N}\right]}$$

$$s^2 = \frac{1}{N-1}\left[\sum X^2 - \frac{(\sum X)^2}{N}\right], \quad s = \sqrt{\frac{1}{N-1}\left[\sum X^2 - \frac{(\sum X)^2}{N}\right]}$$

TABLE 5.1

Computation of σ^2 and σ, and s^2 and s, for two small samples using the definition and computing formulas

Example 1. Distribution 4 (where $\overline{X} = 5.0$)

X	$X - \overline{X}$	$(X - \overline{X})^2$	X	X^2
7	$7 - 5 = 2$	4	7	49
7	$7 - 5 = 2$	4	7	49
6	$6 - 5 = 1$	1	6	36
5	$5 - 5 = 0$	0	5	25
4	$4 - 5 = -1$	1	4	16
4	$4 - 5 = -1$	1	4	16
4	$4 - 5 = -1$	1	4	16
3	$3 - 5 = -2$	4	3	9
		$\sum (X - \overline{X})^2 = 16$	$\sum X = 40$	$\sum X^2 = 216$

A. *Definition formulas*

$$\sigma^2 = \frac{\sum (X - \overline{X})^2}{N} = \frac{16}{8} = 2.00$$

$$\sigma = \sqrt{2.00} = 1.41$$

B. *Computing formulas*

$$\sigma^2 = \frac{1}{N}\left[\sum X^2 - \frac{(\sum X)^2}{N}\right]$$

$$= \frac{1}{8}\left[216 - \frac{(40)^2}{8}\right]$$

$$= \frac{1}{8}(216 - 200)$$

$$= \frac{1}{8}(16) = 2.00$$

$$\sigma = \sqrt{2.00} = 1.41$$

$$s^2 = \frac{\sum (X - \overline{X})^2}{N-1} = \frac{16}{7} = 2.29$$

$$s = \sqrt{2.29} = 1.51$$

$$s^2 = \frac{1}{N-1}\left[\sum X^2 - \frac{(\sum X)^2}{N}\right]$$

$$= \frac{1}{7}\left[216 - \frac{(40)^2}{8}\right]$$

$$= \frac{1}{7}(216 - 200) = \frac{1}{7}(16) = 2.29$$

$$s = \sqrt{2.29} = 1.51$$

table continues

TABLE 5.1 (Continued)

Example 2. Distribution 5 (where $\overline{X}$ = 5.0)

X	$X - \overline{X}$	$(X - \overline{X})^2$	X	X^2
10	10 − 5 = 5	25	10	100
10	10 − 5 = 5	25	10	100
9	9 − 5 = 4	16	9	81
7	7 − 5 = 2	4	7	49
5	5 − 5 = 0	0	5	25
4	4 − 5 = − 1	1	4	16
3	3 − 5 = − 2	4	3	9
2	2 − 5 = − 3	9	2	4
0	0 − 5 = − 5	25	0	0
0	0 − 5 = − 5	25	0	0
		$\sum (X - \overline{X})^2 = 134$	$\sum X = 50$	$\sum X^2 = 384$

A. *Definition formulas*

$$\sigma^2 = \frac{\sum (X - \overline{X})^2}{N} = \frac{134}{10} = 13.40$$

$$\sigma = \sqrt{13.40} = 3.66$$

B. *Computing formulas*

$$\sigma^2 = \frac{1}{N}\left[\sum X^2 - \frac{(\sum X)^2}{N}\right]$$

$$= \frac{1}{10}\left[384 - \frac{(50)^2}{10}\right]$$

$$= \frac{1}{10}(384 - 250)$$

$$= \frac{1}{10}(134) = 13.40$$

$$\sigma = \sqrt{13.40} = 3.66$$

$$s^2 = \frac{\sum (X - \overline{X})^2}{N - 1} = \frac{134}{9} = 14.89$$

$$s = \sqrt{14.89} = 3.86$$

$$s^2 = \frac{1}{N - 1}\left[\sum X^2 - \frac{(\sum X)^2}{N}\right]$$

$$= \frac{1}{9}\left[384 - \frac{(50)^2}{10}\right]$$

$$= \frac{1}{9}(384 - 250) = 14.89$$

$$s = \sqrt{14.89} = 3.86$$

Examples of the use of the computing formulas for σ^2 and σ, and s^2 and s, are shown in Table 5.1.

Computing the standard deviation and variance from a regular frequency distribution. When scores are arranged in the form of a regular frequency distribution, it is convenient to determine the squared deviation (or X^2 in the computing formula) just once for each score value and then multiply it by the frequency of that score value. That is:

Definition Formulas	**Computing Formulas**
$$\sigma^2 = \frac{\sum f(X - \bar{X})^2}{N}$$	$$\sigma^2 = \frac{1}{N}\left[\sum fX^2 - \frac{(\sum fX)^2}{N}\right]$$
$$\sigma = \sqrt{\frac{\sum f(X - \bar{X})^2}{N}}$$	$$\sigma = \sqrt{\frac{1}{N}\left[\sum fX^2 - \frac{(\sum fX)^2}{N}\right]}$$
$$s^2 = \frac{\sum f(X - \bar{X})^2}{N - 1}$$	$$s^2 = \frac{1}{N - 1}\left[\sum fX^2 - \frac{(\sum fX)^2}{N}\right]$$
$$s = \sqrt{\frac{\sum f(X - \bar{X})^2}{N - 1}}$$	$$s = \sqrt{\frac{1}{N - 1}\left[\sum fX^2 - \frac{(\sum fX)^2}{N}\right]}$$

As an illustration, Distribution 4 has been arranged into a regular frequency distribution, and s^2 and s have been computed in Table 5.2. Compare the computations to those for Distribution 4 in Table 5.1 and verify that the results are identical.

TABLE 5.2

Computation of s^2 and s from a regular frequency distribution using the definition and computing formulas

Distribution 4 (where $\bar{X} = 5.0$)

X	f	X − $\bar{X}$	$(X - \bar{X})^2$	$f(X - \bar{X})^2$	X	f	X^2	fX	fX^2
7	2	7 − 5 = 2	4	8	7	2	49	14	98
6	1	6 − 5 = 1	1	1	6	1	36	6	36
5	1	5 − 5 = 0	0	0	5	1	25	5	25
4	3	4 − 5 = −1	1	3	4	3	16	12	48
3	1	3 − 5 = −2	4	4	3	1	9	3	9
2	0	2 − 5 = −3	9	0	2	0	4	0	0
1	0	1 − 5 = −4	16	0	1	0	1	0	0
0	0	0 − 5 = −5	25	0	0	0	0	0	0

$$\sum f(X - \bar{X})^2 = 16 \qquad \sum fX = 40 \qquad \sum fX^2 = 216$$

A. Definition formula

$$s^2 = \frac{\sum f(X - \bar{X})^2}{N - 1} = \frac{16}{7} = 2.29$$

$$s = \sqrt{2.29} = 1.51$$

B. Computing formula

$$s^2 = \frac{1}{N - 1}\left[\sum fX^2 - \frac{(\sum fX)^2}{N}\right]$$

$$= \frac{1}{7}\left[216 - \frac{(40)^2}{8}\right]$$

$$= \frac{1}{7}(216 - 200) = \frac{1}{7}(16) = 2.29$$

$$s = \sqrt{2.29} = 1.51$$

Summary

A second important attribute of a set of scores is its *variability,* or how much the scores *differ from one another.* In the behavioral sciences, this is customarily summarized in a single number by computing the *variance* or its positive square root, the *standard deviation.* The larger the variance or standard deviation, the more different the numbers are from one another. The concept of variability has many practical applications and is particularly important in statistical work.

1. The variance and the standard deviation

1. *Variance*

   Definition formula:

   $$\sigma^2 = \frac{\sum (X - \bar{X})^2}{N}$$

   Computing formula:

   $$\sigma^2 = \frac{1}{N}\left[\sum X^2 - \frac{(\sum X)^2}{N}\right]$$

   Formulas for use with regular frequency distributions:

   $$\sigma^2 = \frac{\sum f(X - \bar{X})^2}{N}$$

   $$\sigma^2 = \frac{1}{N}\left[\sum fX^2 - \frac{(\sum fX)^2}{N}\right]$$

2. *Standard deviation*

   $$\sigma = +\sqrt{\sigma^2}$$

   σ_2 and σ are basic measures of the variability of any set of data.

2. The population variance estimate

When the data of a sample are to be used to estimate the variance of the population from which the sample was drawn, compute the *population variance estimate.*

Definition formula:

$$s^2 = \frac{\sum (X - \bar{X})^2}{N - 1}$$

Computing formula:

$$s^2 = \frac{1}{N-1}\left[\sum X^2 - \frac{(\sum X)^2}{N}\right]$$

Formulas for use with regular frequency distributions:

$$s^2 = \frac{\sum f(X - \bar{X})^2}{N-1}$$

$$s^2 = \frac{1}{N-1}\left[\sum fX^2 - \frac{(\sum fX)^2}{N}\right]$$

Note that in this situation, the standard deviation $s = +\sqrt{s^2}$.

3. The range

Another measure of variability sometimes encountered in the behavioral sciences is the *range*, which is equal to the highest score minus the lowest score. However, the range is at best a rough measure and is generally less useful than the variance or standard deviation.

Chapter 6
Transformed Scores II: Z and T Scores

PREVIEW

Interpreting a Raw Score

What is the advantage of using a transformed score, such as a percentile rank?

When comparing two (or more) scores from different distributions, why is it necessary to refer each score to its mean and standard deviation? What errors are we likely to make if we look only at the raw scores?

Rules for Changing $\overline{X}$ and σ

What will happen to the mean and standard deviation of a set of scores if we do the following: Add a constant to every score? Subtract a constant from every score? Multiply every score by a constant? Divide every score by a constant?

How can these rules be used to obtain transformed scores with *any* desired mean and standard deviation?

Standard Scores (Z Scores)

What is the mean and standard deviation of a set of Z scores? Why is this desirable?

How are Z scores computed?

When raw scores are transformed into Z scores, what happens to the shape of the distribution?

T Scores

Why might we prefer to use T scores instead of Z scores?

How are T scores computed?

What is the mean and standard deviation of a set of T scores?

SAT Scores

What is the mean and standard deviation of a set of SAT scores, and how are these scores computed?

Summary

Appendix: Proofs of Rules for Changing $\overline{X}$ and σ

Interpreting a Raw Score

In Chapter 3, we saw that it can be helpful to transform a raw score into a percentile rank. The percentile rank shows at a glance how the score stands in comparison to a specific reference group.

In Chapters 4 and 5, we identified two important characteristics of a group of scores: its location, frequently summarized by the mean $(\overline{X})$, and its variability, for which we will use the standard deviation (σ). It is possible to deduce how well a given score compares to the reference group by using the mean and standard deviation of that group. As was the case with percentiles, we can build this information into the score itself. That is, we can derive a transformed score that shows at a glance the relationship of the original raw score to the mean, using the standard deviation of the reference group as the unit of measurement.

To illustrate, let us suppose that a college student takes three midterm examinations in three different subjects and obtains the following raw scores:

	English	Mathematics	Psychology
X	80	65	75

On the surface, it might seem as though the student's best score is in English and his poorest score is in mathematics. It would be unwise to jump to such a conclusion, however, since there are several reasons why the raw scores may not be directly comparable. For example, the English examination may have been easy and resulted in many high scores, while the mathematics examination may have been extremely difficult. Or the English examination may have been based on a total of 100 points, and the mathematics examination on a total of only 80 points. The raw scores provide information about the absolute number of points earned, but they give no indication as to how good the performance is, and certainly no indication of how good the performance is compared to others.

Suppose that we specify the mean and standard deviation of each test:

	English	Mathematics	Psychology
X	80	65	75
$\overline{X}$	85	55	60
σ	10	5	15

This additional information changes the picture considerably. Looking at the means, we can see that the scores on the English examination were high, so much so that the score of 80 is below average. On the other hand, both the mathematics and psychology scores are above average. Therefore, the student's poorest result is in English.

The unwary observer might now conclude that the student's best score is in psychology, since that score is 15 points above average while the mathematics score is only 10 points above average. But as we pointed out in Chapter 5, the variability of a distribution of scores also influences the relative standing of a given score. The standard deviation indicates that the "average" variability on the psychology test was 15 points from the mean; some scores were more than 15 units from the mean and some were less. So the student's psychology score of 75, which is 15 points or one standard deviation above average, was exceeded by a number of better scores.* The average variability on the mathematics examination, however, was only five points from the mean. Hence the student's mathematics score of 65 is 10 points or *two* standard deviations above average. It is unusually far above the mean, and is therefore likely to be one of the best scores.†

The picture presented by the raw scores was quite misleading in this instance. It turns out that the student's best score is in mathematics, the next best score is in psychology, and the poorest score is in English, all relative to the other students in each course.

The raw scores of 80, 65, and 75 cannot be compared directly because they come from distributions with different means and different standard deviations. That is, the units in which the raw scores measure are not the same from test to test. This difficulty can be overcome by transforming the scores on each test to a common scale with a specified mean and standard deviation. This new scale will then serve as a "common denominator," enabling us to compare directly the transformed scores of different tests.

Two questions remain: How do we change the scores so that they will have the desired common mean and standard deviation? And, what values of the mean and standard deviation are useful choices for the "common denominator"?

Rules for Changing $\overline{X}$ and σ

Suppose that the mathematics instructor in the previous example suffers a pang of conscience about the low mean, and decides to add five points to everyone's score. Since she has added a *constant* amount to each score (one that is exactly the same for all the scores), she does not have to recompute a new mean via the usual formula; it can readily be proved that adding five

---

* The exact number depends on the shape of the distribution of scores. For example, if the scores are normally distributed, approximately 16% of the scores are more than one standard deviation above the mean (as will be shown in Chapter 9).

† If the distribution is normal, less than $2\frac{1}{2}\%$ of the scores are more than two standard deviations above the mean.

points to everyone's score increases the mean by five points.* Thus, the mean of the transformed scores (symbolized by $\overline{X}_{new}$) will be 60.

In general:

If a constant k is *added* to every score,

$$\overline{X}_{new} = \overline{X}_{old} + k$$

Similarly, it can be proved that:

If a constant k is *subtracted* from every score,

$$\overline{X}_{new} = \overline{X}_{old} - k$$

If every score is *multiplied* by a constant k,

$$\overline{X}_{new} = k\overline{X}_{old}$$

If every score is *divided* by a constant k,

$$\overline{X}_{new} = \overline{X}_{old}/k$$

If the English instructor subtracts 7.5 points from every score, the new mean is 77.5. If he multiplies every score by 4, the new mean is 340. And if he divides every score by 2, the new mean is 42.5. Note that these rules for conveniently determining the new mean work only if every original score is altered by exactly the same amount.

Insofar as the variability of the distribution is concerned, adding a constant to every score or subtracting a constant from every score does *not* change the standard deviation or variance. Adding or subtracting a constant does not change the spread of the distribution, since each score is increased or decreased by the same amount. Insofar as multiplication and division by a constant are concerned, it can be shown that:

If every score is *multiplied* by a positive constant k,

$$\sigma_{new} = k\sigma_{old}$$
$$\sigma_{new}^2 = k^2\sigma_{old}^2$$

If every score is *divided* by a positive constant k,

$$\sigma_{new} = \sigma_{old}/k$$
$$\sigma_{new}^2 = \sigma_{old}^2/k^2$$

---

* The proofs mentioned in this section are presented in the Appendix at the end of this chapter.

If the English instructor adds 10 points or subtracts 6 points from every score, the standard deviation remains 10 (and the variance remains 10^2 or 100). If he multiplies every score by 4, the new standard deviation equals 4×10 or 40 and the new variance equals $4^2 \times 100$ or 1,600. If he divides every score by 2, the new standard deviation equals 10/2 or 5 and the new variance equals $100/2^2$ or 25.

These rules make it possible to obtain transformed scores with any desired mean and standard deviation. For example, scores on the English test can be transformed to scores comparable to those on the mathematics test in two steps:

Procedure	New $\overline{X}$	New σ
1. Divide every score on the English examination by 2.0.	85/2 = 42.5	10/2 = 5
2. Add 12.5 to each of the scores obtained in Step 1.	42.5 + 12.5 = 55	5 (no change)

The first step is to change the standard deviation to the desired value by multiplying or dividing each score by the appropriate constant, which affects the value of both $\overline{X}$ and σ. Then, the desired mean is obtained by adding or subtracting the appropriate constant, which does not cause any further change in σ. When the student's English score of 80 is subjected to these transformations, it becomes (80/2) + 12.5 or 52.5, and it is evident that this score is not nearly as good as the student's mathematics score. Note that both the original and the transformed English scores are half a standard deviation below the means of their respective distributions.

Standard Scores (Z Scores)

The techniques discussed in the preceding section make it possible to switch to any new mean and standard deviation. Therefore, the next logical step is to choose values of $\overline{X}_{new}$ and σ_{new} that facilitate comparisons among the scores. One very useful procedure is to convert the original scores to new scores with a mean of 0 and a standard deviation of 1, called Z scores or *standard scores.*

Standard scores have two major advantages. Since the mean is zero, you can tell at a glance whether a given score is above or below average; an above-average score is positive and a below-average score is negative. Also, since the standard deviation is 1, the numerical size of a standard score indicates *how many standard deviations* above or below average the score is. We saw at the beginning of this chapter that this information offers a valuable clue as to how good the score is; a score one standard deviation above average (that is, a standard score of +1) would demarcate approximately the top 16% in a normal distribution, while a score two standard deviations above average (a standard score of $+2$) would demarcate approximately the top $2\frac{1}{2}$% in a normal distribution.

To convert a set of scores to standard scores, the first step is to subtract the original mean from every score. According to the rules given in the previous section, the new mean is equal to

$$\overline{X}_{new} = \overline{X}_{old} - \overline{X}_{old}$$
$$= 0$$

while the standard deviation is unchanged. Next, each score obtained from the first step is divided by the original standard deviation. As a result,

$$\overline{X}_{new} = \frac{0}{\sigma_{old}} = 0$$

$$\sigma_{new} = \frac{\sigma_{old}}{\sigma_{old}} = 1$$

That is, the mean remains zero, while the standard deviation becomes 1. Summarizing these steps in a single formula gives

$$Z = \frac{X - \overline{X}}{\sigma}$$

where Z is the symbol for a standard score.

Converting each of the original examination scores given at the beginning of this chapter to Z scores yields:

	English	Mathematics	Psychology
X	80	65	75
$\overline{X}$	85	55	60
σ	10	5	15
Z	$\dfrac{80 - 85}{10} = -0.50$	$\dfrac{65 - 55}{5} = +2.00$	$\dfrac{75 - 60}{15} = +1.00$

The standard scores show at a glance that the student was half a standard deviation below the mean in English, two standard deviations above the mean in mathematics, and one standard deviation above the mean in psychology.

When raw scores are transformed into Z scores, *the shape of the distribution remains the same.* We just measure from a new point (the mean instead of zero), with a new unit size (the standard deviation instead of the raw units). To illustrate this point, a set of hypothetical heights in inches for 20 men is presented in Table 6.1 along with the corresponding Z scores. The scores have been arranged in increasing order for clarity. The mean of the raw height scores is 67.80 in. and the standard deviation is 1.78 in., and the Z scores were obtained using the formula given above. For example, the Z score corresponding to 72 in. is equal to

$$Z = \frac{72 - 67.80}{1.78} = +2.36$$

In accordance with the previous discussion, the mean of the Z scores is 0.00 and the standard deviation is 1.00. The raw score and Z score distributions are plotted in Figure 6.1; note that a single graph suffices because the shape of the distribution is the same for both sets of scores. Thus the relationship of the scores to one another is *not* changed by transforming them to standard scores. All that change are the location and the scaling.

Although the height scores are obtained by measuring the distance from the floor to the top of each individual, identical Z scores would be obtained

TABLE 6.1

Hypothetical distribution of 20 height scores expressed as raw scores (in.), Z scores, and raw scores (cm) measured from the top of a 36-in. desk

Person	Height: Raw Score (in.)	Height: Z Score	Height: Raw Score (cm) Measured from Top of 36-in. Desk
1	72	+ 2.36	91.44
2	70	+ 1.24	86.36
3	70	+ 1.24	86.36
4	70	+ 1.24	86.36
5	69	+ 0.67	83.82
6	69	+ 0.67	83.82
7	68	+ 0.11	81.28
8	68	+ 0.11	81.28
9	68	+ 0.11	81.28
10	68	+ 0.11	81.28
11	67	− 0.45	78.74
12	67	− 0.45	78.74
13	67	− 0.45	78.74
14	67	− 0.45	78.74
15	67	− 0.45	78.74
16	67	− 0.45	78.74
17	66	− 1.01	76.20
18	66	− 1.01	76.20
19	66	− 1.01	76.20
20	64	− 2.14	71.12
$\overline{X}$	67.80	0.00	80.77
σ	1.78	1.00	4.52

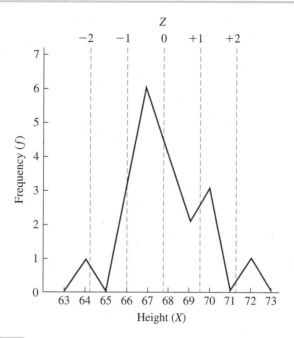

FIGURE 6.1

Frequency distribution of raw scores and *Z* scores in Table 6.1

if every person were measured from the top of a desk that is 36 in. high instead of from the ground. Even a change of scale to centimeters would not affect the *Z* scores, as is shown in the last column of Table 6.1. For example,

in. measured from the floor	*Z* score	cm measured from top of 36-in. desk

$$\frac{70 - 67.80}{.1.78} = +1.24 = \frac{86.36 - 80.77}{4.52}$$

This indicates that *Z* scores give an accurate picture of the standing of each score relative to the reference group no matter where the original scores are measured from, or what scale is used.

Standard scores are used extensively in the behavioral sciences. They also play an important role in statistical inference, as we will see in subsequent chapters.

T Scores

Standard scores have one disadvantage: they are difficult to explain to someone who is not well versed in statistics. A college professor once decided to report the results of an examination as *Z* scores. He was quickly besieged by anxious students who did not understand that a Z score of 0 represents average performance (and not zero correct!), not to mention the agitation of those who received negative scores and wondered how they could ever repay the points they owed the professor.

Since one task of behavioral scientists is to report test scores to people who are not statistically sophisticated, several alternatives to Z scores have been developed. The mean and standard deviation of each such "common denominator" have been chosen so that all of the scores will be positive, and so that the mean and standard deviation will be easy to remember. One such alternative, called T scores, is defined as a set of scores with a mean of 50 and a standard deviation of 10. The T scores are obtained from the following formula:

$$T = 10Z + 50$$

Each raw score is converted to a Z score, each Z score is multiplied by 10, and 50 is added to each resulting score. For example, a height score of 69 in Table 6.1 is converted to a Z score of $+0.67$ by the usual formula. Then, T is equal to $(10)(+0.67) + 50$ or 56.70. It is easy to prove that the above formula does produce the desired mean and standard deviation:

Procedure	New $\overline{X}$	New σ
1. Convert raw scores to Z scores.	0	1
2. Multiply each Z score by 10.	$10 \times 0 = 0$	$10 \times 1 = 10$
3. Add 50 to each score obtained in Step 2.	$0 + 50 = 50$	10 (no change)

Since the mean of T scores is 50, you can still tell at a glance whether a score is above average (it will be greater than 50) or below average (it will be less than 50). Also, you can tell how many standard deviations above or below average a score is. For example, a score of 40 is exactly one standard deviation below average (equivalent to a Z score of -1.00) since the standard deviation of T scores is 10. A negative T score is mathematically possible but virtually never occurs; it would require that a person be over five standard deviations below average—and scores more than three standard deviations above or below the mean almost never occur with real data.

SAT Scores

Scores on some nationally administered examinations, such as the Scholastic Aptitude Test (SAT), the College Entrance Examination Boards, and the Graduate Record Examination, are transformed to a scale with a mean of

500 and a standard deviation of 100. These scores, which we will call SAT scores for want of a better term, are obtained as follows:

$$SAT = 100Z + 500$$

The raw scores are first converted to Z scores, each Z score is multiplied by 100, and 500 is added to each resulting score. The proof that this formula does yield a mean of 500 and a standard deviation of 100 is similar to that involving T scores. (In fact, an SAT score is just ten times a T score.) This explains the apparent mystery of how you can obtain a score of 642 on a test with only several hundred items. And you may well be pleased if you obtain a score of 642, since it is 142 points or 1.42 standard deviations above the mean (and therefore corresponds to a Z score of $+1.42$ and a T score of 64.2).

Summary

In order to compare scores based on different means and standard deviations, it is desirable to convert the raw scores to *transformed scores* with a common mean and standard deviation. Some frequently used transformed scores are Z *scores (standard scores),* with a mean of 0 and a standard deviation of 1; T *scores,* with a mean of 50 and a standard deviation of 10; and *"SAT scores,"* with a mean of 500 and a standard deviation of 100. A transformed score shows at a glance whether a score is above or below average, and how far in terms of standard deviation units, with respect to a specified reference group.

1. General transformations

Operation	Effect on $\overline{X}$	Effect on σ	Effect on σ^2
1. *Add* a constant, k, to every score	new mean = old mean + k	no change	no change
2. *Subtract* a constant, k, from every score	new mean = old mean − k	no change	no change
3. *Multiply* every score by a constant, k	new mean = old mean × k	new σ = old σ × k	new σ^2 = old σ^2 × k^2
4. *Divide* every score by a constant, k	new mean = old mean/k	new σ = old σ/k	new σ^2 = old σ^2/k^2

2. Z scores (standard scores)

$$Z = \frac{X - \overline{X}}{\sigma}$$

3. *T scores*

$$T = 10Z + 50$$

4. *SAT scores*

$$SAT = 100Z + 500$$

$$= 10T$$

Appendix: Proofs of Rules for Changing $\overline{X}$ and σ

Rule 1.   If a constant (positive or negative) k is added to every score,

$$\overline{X}_{new} = \overline{X}_{old} + k$$

PROOF: Each new score may be represented by $X + k$. The mean of these scores, obtained in the usual way by summing the scores and dividing by N, is

$$\overline{X}_{new} = \frac{\Sigma\,(X + k)}{N}$$

$$= \frac{\Sigma\,X + Nk}{N} \qquad \text{(Rule 6, Chapter 1)}$$

$$= \frac{\Sigma\,X}{N} + \frac{Nk}{N}$$

$$= \overline{X}_{old} + k$$

Rule 2.   If a constant (positive or negative) k is added to every score,

$$\sigma^2_{new} = \sigma^2_{old}$$

$$\sigma_{new} = \sigma_{old}$$

PROOF: Each new score may be represented by $X + k$, and the mean of the new scores is $\overline{X}_{old} + k$. Therefore, the definition formula for the variance of the transformed scores becomes

$$\sigma^2_{new} = \frac{\Sigma\,([X + k] - [\overline{X}_{old} + k])^2}{N}$$

$$= \frac{\Sigma\,(X + k - \overline{X}_{old} - k)^2}{N}$$

$$= \frac{\sum (X - \overline{X}_{old})^2}{N}$$

$$= \sigma^2_{old}$$

And since $\sigma^2_{new} = \sigma^2_{old}$, $\sigma_{new} = \sigma_{old}$.

Rule 3. If every score is multiplied by a constant k (positive or negative, and greater than or less than one),

$$\overline{X}_{new} = k\overline{X}_{old}$$

PROOF: Each new score may be represented by kX, and the mean of these scores is

$$\overline{X}_{new} = \frac{\sum kX}{N}$$

$$= k\frac{\sum X}{N} \qquad \text{(Rule 8, Chapter 1)}$$

$$= k\overline{X}_{old}$$

Rule 4. If every score is multiplied by a positive constant k,

$$\sigma^2_{new} = k^2\sigma^2_{old}$$

$$\sigma_{new} = k\sigma_{old}$$

PROOF: Each new score may be represented by kX, and the mean of the new scores is $k\overline{X}_{old}$. The variance is equal to

$$\sigma^2_{new} = \frac{\sum (kX - k\overline{X}_{old})^2}{N}$$

$$= \frac{\sum (k[X - \overline{X}_{old}])^2}{N}$$

$$= \frac{\sum k^2(X - \overline{X}_{old})^2}{N}$$

$$= \frac{k^2 \sum (X - \overline{X}_{old})^2}{N}$$

$$= k^2\sigma^2_{old}$$

$$\sigma_{new} = \sqrt{k^2\sigma^2_{old}}$$

$$= k\sigma_{old}$$

Chapter 7
Additional Techniques for Describing Batches of Data

PREVIEW

Numerical Summaries

Why might the mean and standard deviation *not* be a good way to summarize a particular set of data?

What is a 5-number summary, how is it constructed, and what kinds of information does it provide?

Graphic Summaries

What is a box-and-whisker plot, and how is it constructed?

For what kinds of data are box-and-whisker plots particularly useful?

What is a mean-on-spoke representation, and how is it constructed?

For what kinds of data are mean-on-spoke representations preferable to box-and-whisker plots?

Summary

Thus far, we have discussed various graphic and numerical techniques for describing sets of data. In some situations, pictorial representations will be worth the proverbial thousand words (e.g., histograms; frequency polygons). In others, a tabular format may prove best (e.g., frequency distributions; stem-and-leaf displays). And in still others, a few numbers may be most convenient and effective (e.g., means; medians; standard deviations; Z scores; percentiles). The purpose of descriptive statistics is to summarize a set of data, and the best summary is one that discards the greatest number of minor details while retaining and highlighting important information.

Since the definition of what is important varies from one situation to another, there is no one universally accepted way to describe a set of data. The more techniques you have at your disposal, the better you will be able to choose one that emphasizes those aspects of the data that you deem to be most crucial. Therefore, let's conclude our investigation of descriptive statistics by examining some additional, relatively modern summarization procedures.

Numerical Summaries

As we observed in Chapters 4 and 5, the two most important characteristics of a set of numerical observations are its location and variability. Many distributions are well described by just two numbers: the mean and standard deviation. For example, a class of high-school students' IQ scores might be found to have a mean of 105.2 and a standard deviation of 13.6, conventionally represented as 105.2 ± 13.6.

The kind of distribution that is particularly well described by its mean and standard deviation is one that is symmetric, and has a single mode in the middle of the range of values. This is the "bell-shaped" curve, illustrated in Figure 2.4A and discussed in detail in Chapter 9. Bell-shaped distributions occur frequently in the behavioral sciences, hence the great popularity of the mean and standard deviation as descriptive tools. But some distributions depart substantially from this shape, in which case the mean and standard deviation may be inadequate or even misleading as numerical summaries. (Recall the salary example in Chapter 4.) In such instances, therefore, you will need to seek out more accurate alternatives.

Percentiles and the 5-number summary.   In Chapter 3, we observed that a percentile rank describes a specific score in terms of the proportion of cases falling at or below that score. If a college student's score of 146 on an entrance examination places him at the 80th percentile of the freshman class, then he exceeds (or is equal to) 80% of this group.

In addition to individual scores, entire distributions may be summarized by reporting the score values for various specific percentile ranks (e.g., quartiles or deciles). A particularly effective method using such "order" statistics is the 5-number display devised by Tukey,* who also invented the stem-and-leaf display. The five numbers referred to are the median, the first and third quartiles (i.e., the 25th and 75th percentiles), and the lowest and highest scores. These five numbers are conventionally represented in a box, as follows:

Median		30
First, Third Quartile	22	35
Lowest, Highest Score	9	49

The values of the five numbers are readily obtained from a stem-and-leaf display (see Chapter 3), and they convey a considerable amount of information about the distribution. The ones shown above are from the stem-and-leaf display in Table 2.5, generated from the midterm examination scores in Table 2.3. You can see at a glance that a score of 30 divides the class into top and bottom halves, that the lower half is further divided in half by a score of 22 (the 25th percentile, or first quartile), and that the upper half is further divided in half by a score of 35 (the 75th percentile, or third quartile). The poorest (9) and best (49) scores show you exactly the limits within which the observations fall, and are frequently of interest in their own right.

By providing the limits of the four quarters of the ordered observations, the five numbers convey considerable information about the location, spread, and shape of the distribution. Note for example that the quartiles give the limits of the *middle* 50% of the distribution. That is, the middle 50% of the scores occupy the span from 22 to 35. This also indicates that the middle 50% is not symmetric about the median, since the quarter below the median covers a greater range ($30 - 22 = 8$ score points) than does the quarter above the median ($35 - 30 = 5$ score points). Similar asymmetry is shown by the extreme scores: the bottom half occupies a greater range ($30 - 9 = 21$ points) than does the top half ($49 - 30 = 19$ points).

Now let us compare this distribution with one obtained on the same midterm examination given by a different instructor, whose 5-number summary is:

Median		29
First, Third Quartile	24	33
Lowest, Highest Score	8	48

---

* J. W. Tukey, *Exploratory data analysis* (Reading, Mass.: Addison-Wesley, 1977).

Notice that this distribution does not differ much from the preceding one in location (median of 29 vs. 30), and not at all in overall range (48 − 8 = 49 − 9 = 40). The middle half of this distribution is more symmetric about its median, which is 5 points from the first quartile and 4 points from the third quartile. The most striking difference is that the middle half of the cases occupy much less of the range (33 − 24 = 9 points, compared with 35 − 22 = 13 points for the former distribution). Thus the second distribution is much less variable, even though its overall spread is the same. (This should not be overly surprising, since we noted in Chapter 5 that the range is a relatively poor measure of variability.) This in turn implies that the lowest and highest quarters of the students' scores are each considerably more heterogeneous than the comparable groups in the first instructor's course.

The 5-number summary provides a fairly good indication of the shape of the distribution, as well as giving information about its location and spread, so you can use it with all kinds of data. In fact, the second distribution is approximately bell shaped, while the first is clearly skewed.

Graphic Summaries

Box-and-whisker plots.   The information contained in a 5-number summary may be expressed graphically in a figure called the "box-and-whisker plot." * Although this device may be used to represent a single distribution, it is particularly valuable when two or more distributions are to be compared. Box-and-whisker plots for the two midterm examination distributions discussed above are presented in Figure 7.1, together with their corresponding 5-number summaries.

In a box-and-whisker plot, the middle half of the observations in a given distribution is presented as a vertical rectangle whose top and bottom fall at the third and first quartiles, respectively. The box is divided by a line which represents the median. Then, lines (whiskers) are extended up to the largest and down to the smallest observation. This enables the span and endpoints of each quarter of the observations to be clearly represented, which in turn provides a clear sense of the distribution as a whole.

With the two box-and-whisker plots side by side on the same scale, it is possible to see even more clearly their similarities and differences. This is because it is easier to see and compare lengths of lines than it is to mentally compute differences (and differences between differences). Note in Figure 7.1 how easy it is to perceive the narrower spread of the middle half of the cases in the second class, and the concomitant greater spread of the lowest and highest quarters. Also, with the distributions

---

* Tukey, *op. cit.*

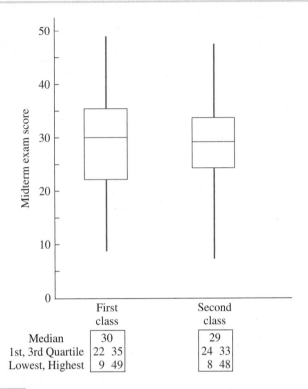

	First class	Second class
Median	30	29
1st, 3rd Quartile	22 35	24 33
Lowest, Highest	9 49	8 48

FIGURE 7.1

Box-and-whisker plots of the midterm examinations for the two classes and their 5-number summaries

summarized in this way, you can better appreciate how similar the two classes are otherwise.

Mean-on-spoke representations. Whereas box-and-whisker plots are particularly useful for irregular or skewed data, a more concise summary can be achieved when a distribution is approximately bell shaped. This graphic representation makes use of the mean and standard deviation, as would be expected in such instances, and is known as the "mean-on-spoke."

To illustrate, consider an experiment wherein two experimental teaching methods (E_1 and E_2) and a control method (C) are used. The three groups are formed by randomly assigning equal numbers of pupils to each one. Following the prescribed number of teaching sessions, all pupils are given an appropriate test to ascertain the effectiveness of each method. Let us assume that the resulting three distributions of scores are approximately bell shaped, and that the means and standard deviations for the three groups are: C: 72 ± 12; E_1: 70 ± 13; E_2: 80 ± 9. With a little effort, you can see that the group taught by experimental method 2 did better on the average than did the other two groups, and had less variability. But you can see this better still if the groups are represented on a graph, with each mean plotted as a point and a spoke running through it that is one standard deviation long on each side. (See Figure 7.2.)

Various implications of bell-shaped distributions can be used to understand the mean-on-spoke representation. For example, in a bell-shaped curve, approximately two-thirds of the observations fall within the limits defined by the mean $\pm$ one standard deviation. Thus, for group E_2 (described by 80 ± 9), you can expect about two-thirds of those pupils to fall between scores of 71 and 89. (How we arrive at two-thirds will become clear when we discuss the characteristics of the normal distribution in Chapter 9.) The graph also shows at a glance that the pupils at the lower end of the spoke in group E_2 perform at about the same level as those at the mean in group C and group E_1.

The methods described in this chapter are descriptive only. You could not use them to draw inferences about whether E_2 is a better teaching method for students in general, rather than just for those who happened to

FIGURE 7.2

Means-on-spokes for the control and two experimental groups and their respective values for mean $\pm$ standard deviation

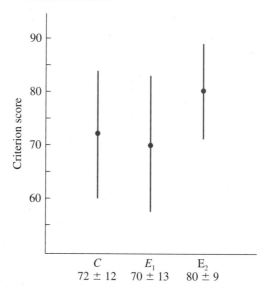

be included in this particular sample.* Such a conclusion would be an important one: if method E_2 is superior, the experiment will have pointed the way toward improving our educational techniques. But if the results obtained from the samples are misleading, and method E_2 will not prove superior when applied to the large population of students, then the researcher will only have pointed the way toward her own embarrassment if she announces a major educational breakthrough. The solution to this dilemma is to perform an analysis using not descriptive statistics, but rather inferential statistics — which is what the rest of this book is about.

Summary

Some additional techniques for describing sets of data were discussed in this chapter. These include:

1. The 5-number summary

This numerical summary provides a good indication of the shape of a distribution by specifying five important values in box form:

	Median	
First Quartile		Third Quartile
Lowest Score		Highest Score

2. Box-and-whisker plots

This graphic summary is particularly valuable when two or more distributions are to be compared. Score values are shown on the vertical axis, with the name of the distribution(s) on the horizontal axis. The middle half of the observations in a given distribution is presented as a vertical rectangle whose top and bottom fall at the third and first quartiles, respectively. The box is divided by a line which represents the median, and lines (whiskers) are extended up to the largest and down to the smallest observation.

---

* Note, however, that with advanced methods, such inferences *can* be made from box-and-whisker plots.

3. Mean-on-spoke representations

This graphic summary is preferable to box-and-whisker plots when a distribution is approximately bell shaped. Here again, score values are shown on the vertical axis; but the mean of a given distribution is plotted as a point, with a vertical spoke running through it that is one standard deviation long on either side.

Part III
Inferential Statistics

Chapter 8

The General Strategy of Inferential Statistics

PREVIEW

Introduction

The typical population of interest to the behavioral science researcher is extremely large; what serious problem does this create? How does the use of samples help to solve this problem?

What new difficulty is created by the use of samples? What type of statistics helps to solve this problem?

The Goals of Inferential Statistics

What are the three major varieties of inferential statistics?

The Strategy of Inferential Statistics

How do we define the probability of a given event?

What is meant by the odds against a given event?

How do we compute the probability that *either* of two events will occur?

How do we compute the probability that one event *and then* another event will occur?

What two hypotheses are made prior to conducting an experiment?

Why must we assume in advance that one of these hypotheses is true? Which hypothesis is assumed to be true? If the results of the statistical analysis point toward this hypothesis, why then must we be very cautious when interpreting the results (as by stating only that we "retain" this hypothesis)? Conversely, why can we be more confident if the results point toward the alternative hypothesis?

Does the use of inferential statistics guarantee that we will reach a correct decision about the population? If not, what then is the advantage of using inferential statistics?

Statistical Models

What is a statistical model? Why is it useful?

Why is it more difficult to determine statistical models in behavioral science research than in simple probability experiments, such as flipping a coin?

(continued next page)

PREVIEW (continued)

What is an experimental sampling distribution? Why is this type of statistical model *not* used in inferential statistics?

What is a theoretical sampling distribution? Why is it important to use such a statistical model (for example, the normal curve model) only in certain situations?

Summary

Introduction

One distressing fact greatly complicates the lives of behavioral scientists: the populations from which they seek to collect data are usually far too large for them to measure every element in the population, or even a sizable portion of the population.

Suppose that a psychologist wishes to test the hypothesis that a new computer software package will improve the performance of students in college-level statistics courses. In this study, the population consists of all college students in the United States who are studying statistics. This is enough to frighten off even the most dedicated researcher! No behavioral scientist can afford the time, effort, and money needed to obtain and analyze data from so many thousands of people.

Therefore, the psychologist might decide to limit the study to 100 statistics students selected in some way from the population. This procedure will provide him with a group that can be measured in its entirety, but a serious problem is created. Since the data are obtained from only a very small part of the population, there is no assurance that these data will accurately reflect what is happening in that population. If this sample of 100 students is atypical of the population as a whole, the psychologist will suffer some unhappy consequences: the results that he publishes, and his decision concerning the merits of the software package, will subsequently be contradicted by other researchers who try to verify his findings with different samples of students drawn from this population. When it is discovered that the psychologist's research findings were incorrect, he will suffer professional loss of face. More importantly, the effort to advance our scientific knowledge in this area will undergo a setback while researchers try to untangle the various conflicting results. Unfortunately, the only sure way to prevent this calamity would be to measure the entire population—which, as we noted above, is impossible.

In sum: Behavioral scientists need to draw conclusions about populations, but they usually can measure only a small part of those populations. This problem affects all areas of the behavioral sciences. For example, an industrial psychologist studying the effects of pay on job satisfaction cannot measure the entire population of all paid employees in the United States. Even if he were to restrict his attention to one type of job, the population would still be far too large to measure in its entirety. An experimental psychologist studying the performance of rats in a maze under varying conditions cannot run all laboratory rats in the world in her experiment, nor can she obtain all possible runs from any one rat. An anthropologist interested in the effects of different cultures on motivation in children cannot study all of the children from each of the cultures in which she is interested. In fact, this predicament is so common that most behavioral science research would not be possible without some effective resolution. As you may have guessed from the title of this chapter, the answer lies in the use of *inferential statistics* — techniques for drawing inferences about an entire population, based on data obtained from a sample drawn from that population.

The Goals of Inferential Statistics

There are three kinds of inferences that the behavioral scientist wishes to make about a population. One useful procedure is to estimate the values of population parameters. Given a sample whose mean and variance have been computed, we may wish to estimate the mean and variance of the parent population. Any statistic computed from a single sample (such as $\overline{X}$ or s) that provides an estimate of the corresponding population parameter (such as μ or σ) is called a *point estimate.*

For some purposes, a point estimate is not sufficiently informative, since it leaves us with no idea of how far off from the parameter it may be. Such information is supplied by an *interval estimate,* a range of values that has a known probability of including the true value of the parameter. For example, suppose that a sample of white mice runs through a maze in an average of 60 seconds. A point estimate would state that the most reasonable value of μ is 60. If an estimate in terms of a single number is *not* essential, the researcher may decide to use appropriate techniques of inferential statistics to determine an *interval* that is likely to include the population mean. Suppose she finds this interval to be 54–66, or 60 ± 6. This interval estimate divides all numerically possible values of μ into two sets: "likely" (54–66) and "unlikely" (less than 54 and more than 66), with "unlikely" having a precise meaning to be discussed in subsequent sections.

A third important use of inferential statistics is to assess the probability of obtaining certain kinds of sample results under certain population conditions. The aforementioned psychologist might wish to determine the probability that a sample of 100 statistics students, who have been switched to computer learning techniques and have gained an average of 5 points in their examination scores, come from a population where the mean gain in examination scores is zero. In other words, if the computer learning techniques do *not* work on the average for the *population,* how likely are we to get a *sample* of 100 statistics students that improved by 5 points?

The Strategy of Inferential Statistics

To illustrate the general strategy of inferential statistics, let us look at an example, based on games of chance, that has a great many similarities to the behavioral research paradigm. To keep the computations as simple as possible, we will choose a straightforward game—that of flipping coins with a friend. Suppose that you wager $1 on the result of the toss of a coin, your friend calls "heads," and the coin lands heads up. You are now out one dollar, but resolve to recoup your loss on the next toss. Unfortunately, your friend continues to call "heads," the coin lands heads up six times in a row, and you lose six consecutive times.

At this point, you begin to entertain some suspicions about the possibility of a biased coin (especially if your friend insists on being the one to call the toss). If the coin is honest, you would like to continue to play and try to regain some of the $6 you have lost; but if your friend has foisted off a biased coin on you, you should terminate the game and denounce your friend as dishonest. For simplicity, we will assume that other possible alternatives, such as mutual agreement to use a new coin, have been rejected by the contestants. We can summarize these possibilities in the form of experimental hypotheses:

HYPOTHESIS 0:   The results are due to chance. The coin is honest, and your friend has had a run of luck.

HYPOTHESIS 1:   The coin is biased (in favor of falling heads up).

It makes a difference as to what decision you choose, as can be seen from the following table comparing your decision with what the true state of affairs might be:

		true state of affairs or state of the population	
		Hypothesis 0 Is True; Results Are Due Solely to Chance	Hypothesis 1 Is True; Coin Is Biased
your decision	do not reject Hypothesis 0 (keep playing)	correct decision	error—continued financial support of dishonest friend
	reject Hypothesis 0 (quit game and denounce friend)	error—unfair slander of honest friend	correct decision

The only way to ensure a correct decision in this touchy situation would be to measure the entire population. In order to do this, it would be necessary to flip the coin an infinite number of times. If the coin were ultimately found to fall heads up more than half the time, your friend would have an unfair advantage, while if the coin came up heads half the time and tails half the time, the game would be fair. Unfortunately, even if generations of people did nothing but flip the accursed coin during every waking moment of their lives, the task of registering an infinite number of tosses would not be finished until the end of time — so you can hardly afford to wait around until the results of this "experiment" are in.

A better procedure is to draw a sample of tosses from the population. In practice, a fairly large sample should be obtained (perhaps as many as 100 flips of the coin, if your willpower survives that long). For purposes of this illustrative example, let us assume that you must reach a decision based on the six results you have already observed. How can you use the information from this sample to draw inferences about whether or not the coin is biased? In order to answer this question, you must first learn some basic definitions and computations regarding the probability of a given event.

Probability

Definition of probability.   The *probability* (*P*) of a given event is defined by the following fraction:

$$P(\text{event}) = \frac{\text{number of ways the specified event can occur}}{\text{total number of possible events}}$$

For example, if an unbiased coin is flipped and you wish to determine the probability that heads will come up, the solution is

$$P(\text{H}) = \frac{1}{2} = \frac{1 \text{ head}}{2 \text{ possible events: head or tail}}$$

Similarly, the probability of "tails up" is also $\frac{1}{2}$. Decimals are often used to express probabilities, so we can also write $P(\text{H})$ (that is, the probability of obtaining a head on one toss of the coin), or $P(\text{T})$, as .50. These numbers express the fact that if the coin is fair, in the long run heads will occur half the time and tails will occur half the time.

To provide some further illustrations, let us assume that you have a standard 52-card deck of playing cards. The deck contains 52 cards divided into four suits (spades, hearts, diamonds, and clubs), each of which contains thirteen cards (two through ten, jack, queen, king, and ace). The deck is thoroughly shuffled, and one card is drawn. The probability that it is the king of diamonds is 1/52 (or .019); there is only one way for that event to occur (only one king of diamonds) and 52 possible equally likely events (cards that could be drawn). In fact, the probability of drawing any one specific card from the deck is 1/52. The probability of drawing a nine is 4/52 or 1/13 (or .077), since there are four ways for this event to occur (nine of spades, nine of hearts, nine of diamonds, and nine of clubs) out of the total possible 52 events. If we define a "spot card" as anything below a ten, the probability of drawing a spot card in clubs is 8/52, or 2/13 (or .15), for there are eight specific equally likely events (the two, three, four, five, six, seven, eight, and nine of clubs) in the 52-card deck.

Probability cannot be less than 0 or greater than 1. If an event has a probability of 1, that means that it must happen. The probability of drawing either a red or a black card from a standard deck is 1.00, since all of the cards in the deck are either red or black (that is, $P = 52/52 = 1.00$). The probability of drawing a purple card with yellow polka dots is zero ($P = 0/52 = 0$).

Odds.   The odds against an event are defined as the ratio of the number of unfavorable outcomes to the number of favorable outcomes, all outcomes being equally likely.

EXAMPLE:

Outcome	Total Events (T)	Number of Ways of Obtaining the Event (W)	Number of Ways of Not Obtaining the Event (T − W)	Probability of Event (W/T)	Odds Against Event (T − W to W)
king of diamonds	52	1	51	1/52	51 to 1
a nine	52	4	48	4/52 or 1/13	48 to 4 or 12 to 1
spot card in clubs	52	8	44	8/52 or 2/13	44 to 8 or 11 to 2 ($5\frac{1}{2}$ to 1)

If the odds against your winning a game are (say) 12 to 1, you need to collect 12 times your wager when you win for the game to be fair. If you stand to win more than 12 times your bet, it is to your advantage to play the game. If you collect less than 12 times your bet when you win, you should refuse to play because you figure to lose in the long run.

The probability of A or B.   The probability of *either of two events* taking place is computed by using the following formula:

$$P(A \text{ or } B) = P(A) + P(B) - P(A \text{ and } B)$$

EXAMPLE: Suppose that you will win a prize if you draw *either* a king *or* a club from a standard 52-card deck in a single try. The probability of winning is equal to

$$P(\text{king}) + P(\text{club}) - P(\text{king and club})$$

$$= 4/52 + 13/52 - 1/52$$

$$= 16/52 \text{ (or .31)}$$

To verify the above calculations, let us tackle the problem the long way. There are the usual 52 total possible events, and the number of favorable events may be counted as follows:

King (4 Favorable Events)	Club (13 Favorable Events)
king of spades	ace of clubs
king of hearts	*king of clubs*
king of diamonds	queen of clubs
king of clubs	jack of clubs
	ten of clubs
	nine of clubs
	eight of clubs
	seven of clubs
	six of clubs
	five of clubs
	four of clubs
	three of clubs
	two of clubs

There appear to be 17 favorable events, 13 clubs and 4 kings. But the king of clubs has been counted twice, and there is only one king of clubs in the deck. Therefore, the probability of king *and* club (1/52) must be subtracted, and the resulting probability (16/52) shows that there are 16 separate and distinct favorable events (cards) out of the total of 52 in the deck.

If events A and B cannot happen simultaneously (in mathematical terminology, are *mutually exclusive*), then $P(A \text{ and } B) = 0$. For example, the probability of drawing a king or a queen from the deck is equal to $4/52 + 4/52 - 0/52 = 8/52$. The events are mutually exclusive because no card in the deck is *both* a king and a queen.

The probability of A and then B.   Suppose that a card is drawn from a standard deck, looked at, and replaced in the deck. The deck is shuffled thoroughly, and a second card is drawn. We wish to know the probability of obtaining a king on the first draw *and then* the ace of spades on the second draw. We assume that the two draws are *independent* — that is, what happens on the first draw does not affect the probabilities on the second draw (because of the shuffling). The solution is given by the formula

$$P(A \text{ and then } B) = P(A) \times P(B)$$

In the present example, the answer is

$$P(\text{king and then ace of spades}) = P(\text{king}) \times P(\text{ace of spades})$$

$$= 4/52 \times 1/52$$

$$= 4/2704 \text{ (or .0015)}$$

If the card drawn on the first try is *not* returned to the deck, the probability becomes $4/52 \times 1/51$ (or $4/2652$). Under this procedure, only 51 cards remain in the deck after the first draw, and the probability of success on the second draw is therefore $1/51$.

To return to the coin problem, the probability of obtaining a head on each of two consecutive flips of a coin is $P(\text{H}) \times P(\text{H})$, which is equal to $1/2 \times 1/2$ or $1/4$. This can be verified by listing all possible outcomes of two flips of the coin:

First Flip	Second Flip	Overall Result	Probability
H	H	2 heads	1/4 (or .25)
H	T	1 head, 1 tail	1/4
T	H	1 head, 1 tail	1/4
T	T	2 tails	1/4

Of the four possible outcomes (all of equal probability), only one yields two heads, so this probability is $1/4$. Similarly, the probability of two tails is $1/4$, and the probability of obtaining one head and one tail irrespective of order is $2/4$ or $1/2$. Since the events are independent (that is, the results of the first flip do not affect the probabilities on the second flip), the probabilities are exactly the same if we flip two coins once.

Conclusion of the Coin "Experiment"

Much more could be said about probability, but we will leave further discussion of this interesting area for the mathematicians and return to the problem concerning flipping coins with a friend. In your quest to determine whether or not the coin is biased, it would be helpful to know the probability of obtaining six consecutive heads. You would have little reason to accuse your friend of unethical practices if this were a common, everyday occurrence. On the other hand, if six heads in a row were a "million-to-one" shot ($P = 1/1,000,000$ or $.000001$), you should be skeptical about the reason for your losses. The best strategy in such a case might well be to assume that bias in the coin, rather than a spectacularly lucky event, was responsible for the results.

If you have been thinking ahead, you may have observed that the rule for determining $P(A \text{ and then } B)$ would appear to yield information that would be helpful in the present dilemma. That is, you might wish to make use of the fact that

$$P(\text{six consecutive heads}) = P(\text{head on flip \#1}) \times P(\text{head on flip \#2})$$
$$\times P(\text{head on flip \#3}) \times P(\text{head on flip \#4})$$
$$\times P(\text{head on flip \#5}) \times P(\text{head on flip \#6})$$

But this formula requires that you know the probability of a head on any one flip, and this probability is equal to 1/2 *only if the coin is unbiased.* If the coin is "loaded," the true probability of a head might be anything other than 1/2; $P(H)$ could equal 2/3, 3/4, or even 1 (if the coin were two-headed). You would like to know the probability of obtaining six consecutive heads in order to determine whether the coin is biased—that is, more likely to fall heads up on any given trial. However, you need to know the probability that the coin will fall heads up on a given trial to determine the probability of six consecutive heads.

To extricate yourself from this predicament, you must use the only certain fact at your disposal: *if the coin is fair, the probability of a head on one trial is 1/2.* This is done as follows:

Step 1. Assume that the coin is fair (that $P(H) = 1/2$).

Step 2. Determine the probability of six consecutive heads, basing the calculations on the assumption stated in Step 1.

Step 3. The calculations in Step 2 yield the probability of six consecutive heads *if* the coin is fair. You may then choose one of two decisions:

Either

A. Based on the probability computed in Step 2, you decide *not* to reject the assumption that the coin is fair;

or

B. The probability computed in Step 2 is so small that the assumption in Step 1 is best regarded as incorrect. While such an event could occur, it is so unlikely that it is better to abandon the assumption that $P(H) = 1/2$. So you reject this assumption, and conclude that the coin is *not* fair.

You can now complete the experiment. *Assuming that the coin is fair,* the probability of six consecutive heads is equal to $(1/2)^6$ or

$$1/2 \times 1/2 \times 1/2 \times 1/2 \times 1/2 \times 1/2 = 1/64 \quad \text{or} \quad .016$$

Your final task is to choose between two decisions. Either the coin is fair, and an event with probability .016 has occurred (Hypothesis 0). *Or* the coin is unfair, $P(H)$ *is not equal to 1/2, and therefore P (six consecutive heads) is not equal to .016* (Hypothesis 1). Remember that the value of .016 is the correct probability for six consecutive heads only if the coin is fair and $P(H) = 1/2$. If this assumption is abandoned, the probability of six heads in a row is *not* necessarily .016. For example, if the coin is so drastically biased that it always falls head up (as would happen if it were a two-headed coin), $P(H) = 1.00$ and $P(\text{six consecutive heads}) = (1.00)^6 = 1.00$. Less drastically, if for that coin $P(H) = .75$, then $P(\text{six consecutive heads}) = (.75)^6 = .18$.

Behavioral scientists commonly follow the arbitrary (but reasonable) practice of regarding events with probabilities less than .05 as "sufficiently unlikely to justify rejecting Hypothesis 0." Events with probabilities greater than .05 are "not sufficiently unlikely" to justify such rejection.* Using this rule, you would abandon the assumption that the coin is fair and terminate the game. Rather than assume that an "unlikely" event had occurred (six consecutive heads with a fair coin), you would instead conclude that the coin was biased in favor of falling heads up.

Now that you have reached a decision, the "experiment" is completed. The statistical analysis cannot guarantee a correct decision (remember, you cannot ensure that you are right because you are unable to measure the whole population), and it is still possible that you have made an error (here, that you are being unfair to a friend who is actually honest). However, statistics will point the way to the "best bet"—the decision about the population that is most warranted by the results obtained from the sample.

The Coin Experiment and Behavioral Science Research Compared

The analysis of a behavioral experiment by means of inferential statistics proceeds in an almost identical fashion to the analysis involving the flipping of the coins. If you understand the rationale underlying the coin experiment, you should have little trouble understanding the use of inferential statistics in the behavioral research experiment.

Let us return to the psychologist mentioned at the outset of this chapter, who hopes to demonstrate gains in students' knowledge of statistics if a new computer software package is used. Since it is impossible to measure the entire population, he obtains a sample of 100 statistics students, chosen in such a way as to avoid bias insofar as possible. The students are given the computer learning materials, and a 5-point increase in the average examination score of the group results. The psychologist then applies the appropriate inferential statistic. He finds to his regret that the probability of an increase this great (or more), assuming that this computerized method does *not* work, is .24. He therefore does not reject Hypothesis 0. A 5-point or greater increase in average examination scores in the sample *is* reasonably likely to occur if the mean increase in the population is zero and the new software package doesn't work.

The parallel between the coin experiment and the behavioral research paradigm is summarized in Table 8.1.

The techniques discussed in the remainder of this book are procedures for computing (a) point estimates, (b) interval estimates, and (c) probability

---

* Since this is a matter of convention and is not based on any mathematical principle, common sense is essential in borderline cases. If, for example, $P = .06$, the technically correct procedure is to call the event "not sufficiently unlikely," but .06 is so close to .05 that the best plan is to repeat the experiment and gain additional evidence. The reason that a guide such as the ".05 rule" is needed is that *you must reach a decision*—either to keep playing or to stop the game. Also, some researchers use a more stringent criterion for a judgment of "unlikely," namely .01; and others use these two and other criteria (such as .10) flexibly, depending on the circumstances of the research.

TABLE 8.1

Coin experiment and behavioral research compared

	Coin Experiment	Behavioral Research (Computer Learning Experiment)
Population	Infinite number of flips of coin	Vast number of people (statistics students)
Sample	Smaller, more feasible number of flips	Smaller, more feasible number of people
Hypothesis 0	Assume that coin is fair, and that any deviation from exact 50–50 heads-tails split in sample is due to chance	Assume that this software package is *not* useful, and that any deviation from exact zero improvement in sample is due to the particular cases that happened to fall in the sample, that is, chance
Hypothesis 1	Results are *not* due to chance (coin is biased)	Results are *not* due to chance (this software package *is* useful)
Experiment	Obtain data (flip coin 6 times — result: 6 heads)	Obtain data (measure sample — result: 5-point mean increase)
Statistical analysis	Compute P (six consecutive heads)	Compute P (at least a five-point increase)
Results of statistical analysis	P (six consecutive heads) *if* coin is fair (answer: .016)	P (at least a five-point increase) *if* this software package is *not* beneficial (answer: .24)
Decision if P is "not sufficiently small" using ".05 rule" ($P = .051 – 1.00$)	Do *not* reject Hypothesis 0. The assumption that the coin is fair is a reasonable one. Continue playing. *Risk:* Continued financial support of dishonest friend. (Size of risk not conveniently determined. See Chapter 14.)	Do *not* reject Hypothesis 0. This assumption that this software package is *not* useful is a reasonable one. Retain standard teaching methods. *Risk:* Failure to adopt new method that is actually better. (Size of risk not conveniently determined. See Chapter 14.)

table continues

values such as the .24 in the hypothetical computer learning experiment. The reason that the rest of this book has so many pages is that different kinds of inferential statistics are needed for different kinds of experiments designed to answer different kinds of questions.

TABLE 8.1 (Continued)

	Coin Experiment	Behavioral Research (Computer Learning Experiment)
Decision if P is "sufficiently small" using ".05 rule" ($P = .00 - .05$)	Reject hypothesis 0; accept Hypothesis 1. The assumption that the coin is fair is *not* reasonable. Terminate game. *Risk:* Unfair slander of honest friend (size of risk $\leq P$).	Reject Hypothesis 0; accept Hypothesis 1. The assumption that this software package is not beneficial is *not* reasonable. Switch to computer learning. *Risk:* Adoption of new method that is no better or worse than standard teaching methods (size of risk $\leq P$).
Actual decision	Reject Hypothesis 0. Terminate game. Since hypothesis 0 has been rejected, we no longer need assume that an event with probability .016 has occurred.	Do not reject Hypothesis 0. It cannot be concluded that this software package is beneficial under these circumstances.

Statistical Models

Suppose that in the coin experiment, *five* heads had been obtained in six flips (and your friend is therefore ahead $4). What decision should be reached about the possibility of a biased coin in this case? To compute this probability, it is necessary to enumerate all possible ways of obtaining five heads in six tosses.*

Possible outcomes						
Flip:	1	2	3	4	5	6
A	H	H	H	H	H	T
B	H	H	H	H	T	H
C	H	H	H	T	H	H
D	H	H	T	H	H	H
E	H	T	H	H	H	H
F	T	H	H	H	H	H

As you may have guessed from the fact that the probability of six consecutive heads is 1/64, there are a total of 64 possible and equally likely out-

---

* Mathematics makes use of special techniques, called "permutations and combinations," to enumerate the various possibilities. Since a knowledge of such mathematics is not necessary for comprehension of the material in this book, it will be omitted.

comes if a fair coin is flipped six times. Thus, the probability of any one of the outcomes shown above is 1/64, and the total probability of five heads in six flips is equal to

$$P(A \text{ or } B \text{ or } C \text{ or } D \text{ or } E \text{ or } F)$$

Since these events are mutually exclusive, this equals 1/64 + 1/64 + 1/64 + 1/64 + 1/64 + 1/64 = 6/64 or .094.

Using the same procedures, we can compute the probability of any number of heads if a coin is flipped six times. The results are as follows:

Number of Heads	Probability
0	1/64 = .016
1	6/64 = .094
2	15/64 = .234
3	20/64 = .312
4	15/64 = .234
5	6/64 = .094
6	1/64 = .016
	64/64 = 1.000

This *statistical model* gives the probabilities of events (number of heads) under stated conditions (that $P(H) = 1/2$). If an observed event or sample result is highly unlikely under the model, then the assumptions about the conditions may need revision. When you obtained six heads in the previous experiment, this result was very unlikely in terms of the statistical model ($P = .016$). You therefore concluded that the assumption underlying the model (that $P(H) = 1/2$) was incorrect — that is, that the coin was biased. If instead you obtained five heads in six flips, you would *not* reject the assumption that the coin was fair. The probability of obtaining five or more heads if $P(H) = 1/2$ is equal to $P(5 \text{ heads or } 6 \text{ heads})$, which is equal to $P(5 \text{ heads}) + P(6 \text{ heads})$ or .094 + .016 or .11, and this is a "not sufficiently unlikely" result in terms of the ".05 rule." It is particularly "not sufficiently unlikely" if the question is, "What is the probability of 5 or more heads *or* 5 or more tails?" — for which the answer is .11 + .11 = .22. Note that the model presents information about any other possible outcome. We could also define the models for games involving different numbers of flips in a similar fashion.

Unfortunately, it is more difficult to determine statistical models in behavioral science research. In the coin experiment, there is one favorable event (head) and a total of two possible and equally likely events (head or tail), so the probability of obtaining a head on one flip of a fair coin is easily defined as 1/2. But what of the computer learning experiment and the 5-point increase in examination scores? Defining "favorable events" and "total possible events" in this situation — or in any research experiment — is by no means a simple task, yet a statistical model is essential in order to provide the probability values that the experimenter needs to reach a decision.

One possible (but not practical) way to obtain the necessary statistical model would be to draw a large number of samples of the same size from the defined population, and determine on an empirical basis how often the various alternatives occur. The samples should be *random*—that is, drawn in such a way as to minimize bias and make the sample typical of the population insofar as possible. A *random sample* is defined as a sample drawn in such a way as to (1) give each element in the population an equal chance of being drawn *and* (2) make all possible samples of that size equally likely to occur. If the mean of each sample is computed and a frequency distribution of the many means is plotted, we will have an idea how often to expect a sample with a particular mean. Such a frequency distribution is called an *experimental sampling distribution* because it is obtained from observed or experimental data.* Note that this is not how distributions of sample values are ordinarily determined, but merely how one *might* determine them.

TABLE 8.2

Hypothetical frequency distribution of mean height for 1,000 samples (sample $N = 100$)

(A) Mean Height (in.)	(B) Number of Samples (f)	(C) Proportion of Samples ($p = f/N$)
68.55 or more	0	.000
68.50–68.54	1	.001
68.45–68.49	1	.001
68.40–68.44	4	.004
68.35–68.39	11	.011
68.30–68.34	19	.019
68.25–68.29	31	.031
68.20–68.24	56	.056
68.15–68.19	81	.081
68.10–68.14	112	.112
68.05–68.09	116	.116
68.00–68.04	141	.141
67.95–67.99	131	.131
67.90–67.94	103	.103
67.85–67.89	76	.076
67.80–67.84	57	.057
67.75–67.79	28	.028
67.70–67.74	21	.021
67.65–67.69	7	.007
67.60–67.64	2	.002
67.55–67.59	2	.002
67.54 or less	0	.000
Total	1,000	1.000

---

* It is also sometimes called a Monte Carlo distribution, particularly when generated by means of an electronic computer.

As an example, suppose that we draw a random sample of 100 American men from the population of American men, measure the height of each man in the sample, and compute the mean height of the sample. We then replace the sample into the population and draw another random sample of 100 American men, and compute the mean height of this sample. In a startling display of perseverance, we obtain a total of 1,000 samples (each of $N = 100$). Table 8.2 shows the frequency distribution of mean heights of the 1,000 samples, and Figure 8.1 graphically represents the frequency distribution.

The experimental sampling distribution serves as the statistical model for assigning probabilities to any mean "height" observation, and Column C of Table 8.2 serves the same function as the table of probabilities of any number of heads in six flips of a coin. The probability of obtaining a sample of a particular mean height is thus defined as the number of *empirically determined* favorable events to the number of *empirically determined* total events. If in 1,000 samples, a mean height in the interval 67.90–67.94 occurs 103 times, then the probability that *any one* sample of size $N = 100$ will have a mean height between 67.90 and 67.94 is 103/1,000 or .103. (Also, if we were to obtain many other random samples of $N = 100$ from this population, we would expect approximately 10% of the samples to have a mean between 67.90 and 67.94.) The probability of drawing a sample with a mean height of 68.25 *or more* would equal .031 + .019 + .011 + .004 + .001 + .001, or .067. Other probabilities could be determined in a similar fashion.

FIGURE 8.1

Hypothetical distribution of mean height for 1,000 samples (sample $N = 100$)

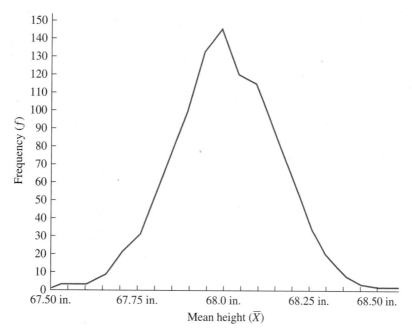

The above procedure for determining an experimental sampling distribution is so laborious that it hardly represents any gain at all over measuring the entire population; it was presented so that you could visualize and understand the result. In practice, you will use *theoretical sampling distributions.* In many instances you will be able to benefit from work done by previous researchers and statisticians, and the form of the sampling distribution of the variable in which you are interested will be known in advance. Observe that in the experimental sampling distribution of height, most of the sample means cluster about the grand mean of 68.00, and there are fewer and fewer observations as one goes further away from the grand mean. This distribution closely approximates the theoretical *normal curve,* a statistical model that describes the sampling distributions of several statistics of interest, including means. Use of the theoretical normal curve model in the proper situations will enable you to avoid the toil and trouble of determining an experimental sampling distribution in order to ascertain the needed probabilities.

For example, the sampling distribution of means is known to have the shape of a normal curve for large samples. So you can proceed on the assumption that if you were to take a great many random samples of a given size from the population, the means of the samples would be normally distributed about the population means (μ), just as they were in the height illustration. By making this assumption, probabilities are assigned on the basis of the theoretical sampling distribution, and an experimental or observed distribution of sample means is no longer necessary. You can then draw just one sample and compare the results to the theoretical model. In addition to means, this strategy can also be used with other statistics, as we will see. The major precaution that should be emphasized is that you must be careful to use the normal curve model only in situations where it is proper to do so, for the normal curve model is not the only model used to make inferences about the population. It *is* one of the most important, however, and the one with which the discussion of specific procedures in inferential statistics will begin.

Summary

1. Probability

Definition:

$$P = \frac{\text{number of ways the specified event can occur}}{\text{total number of possible events}}$$

Odds against an event = number of unfavorable events to number of favorable events

Additive law:

$$P(A \text{ or } B) = P(A) + P(B) - P(A \text{ and } B)$$

Multiplicative law:

If A and B are independent events,
$$P(A \text{ and then } B) = P(A) \times P(B)$$

2. The general strategy of inferential statistics

Behavioral scientists usually wish to *draw inferences about a population* that is too large to measure in its entirely, *based on information obtained from a sample* drawn *randomly* from that population. An initial hypothesis, called "Hypothesis 0," is assumed to be true so that the statistical calculations can be performed; these calculations yield the probability of obtaining the observed result *if* Hypothesis 0 is true. Using this probability, Hypothesis 0 is *rejected* in favor of the alternative hypothesis (Hypothesis 1) if the results are "sufficiently unlikely" to occur if Hypothesis 0 is true. Hypothesis 0 is *retained* (*not rejected*) if the results are "not sufficiently unlikely" to occur if Hypothesis 0 is true. A summary of such probabilities for all possible outcomes in a given experimental situation is called a *statistical model.* Since it is not feasible to develop such models empirically, *theoretical* models are used to decide whether to retain or reject Hypothesis 0.

Chapter 9

The Normal Curve Model

PREVIEW

Introduction

Why is the normal distribution a *theoretical* sampling distribution?

What are two reasons why the normal curve model is useful?

Score Distributions

How is the total population depicted in a graph? As a probability?

In a graph, how do we depict the proportion of the population with scores between two specified values?

We would like to determine the numerical value of such proportions (for example, that 46% of the population of college students scores between 500 and 675 on the SAT). A mathematician would use integral calculus. How does the use of standard scores and the normal curve table replace the need for calculus?

How is the normal curve table used?

What percent of the area under the normal curve falls between the mean and one and two standard deviation units?

Characteristics of the Normal Curve

Where in the normal curve are values more likely to occur?

What is the shape of the normal curve? Is it unimodal? Is it symmetric?

Illustrative Examples

How do we calculate the proportion of the population that can be expected to have scores between two specified values, or above or below one specified value?

How do we calculate the score that is required to be within a specified proportion of the population (for example, the top 10%)?

Summary

Introduction

Having taken pains to justify the existence of statistical models, it is now time to examine one in detail. A statistical model that you will find extremely useful in behavioral research is the *normal distribution*. This model is a *theoretical* distribution, since its shape is defined by a mathematical equation* and would *never* be exactly duplicated by empirical data. There are several reasons why this model is so important:

1. For many variables, you can legitimately *assume* that the shape of the *population distribution* is approximately normal. That is, *without measuring the population,* you can assume the score distribution to be more or less as shown in Figure 9.1.

2. Most importantly, the *sampling* distribution of various statistics tends to be normal if the sample sizes are large. Suppose that a great many random samples are drawn from a population; all are of the same size and contain at least 30 observations. If you compute the mean of each sample

FIGURE 9.1

The normal curve: a theoretical distribution

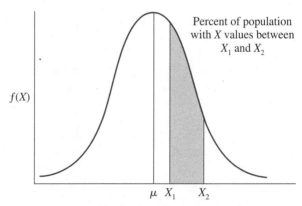

Percent of population with X values between X_1 and X_2

$f(X)$

μ X_1 X_2

X value (may be raw scores or the sampling distribution of a statistic such as means)

---

* The formula for the normal curve is fairly complicated:

$$f(X) = \frac{1}{\sigma\sqrt{2\pi}}\, e^{-(X-\mu)^2/2\sigma^2}$$

where $\pi = 3.1416$ and $e = 2.7183$, both approximately.

and display all of these means in a frequency distribution, the shape of this distribution will be approximately normal.*

Thus, Figure 9.1 may be used to represent either (1) individual *scores* in a normally distributed population (for example, the height of every adult male in the United States), or (2) a normal *sampling distribution* of a statistic (for example, the *mean height* of each of 1,000 large samples). This chapter deals with the normal distribution of individual scores in a population; drawing inferences about the mean of a population will be discussed in the next chapter.

Score Distributions

As with any frequency distribution, the horizontal axis of the normal distribution in Figure 9.1 represents score values and the vertical axis is related to frequency. Also, all values in Figure 9.1 are located *under* the curve. So the *total population* is expressed in the graph as the *total area under the normal curve*, and we can arbitrarily define this area as 100% without bothering about the exact number of cases in the population. Alternatively, we may move the decimal point two places to the left and define the total area under the normal curve as 1.00. This definition is useful when we wish to answer questions that deal with probability, since the sum of all the specific probability values in any statistical model is 1.00 (as we observed in the preceding chapter).

Now suppose that you are asked to find the proportion of the population with values between points X_1 and X_2 in Figure 9.1. Graphically, the answer is given by the percent *area between* these two points. This is equivalent to the percent frequency of these values, and/or to the probability of occurrence of these values.

If you have taken an elementary calculus course, you know that the numerical value of the area between these two X values may be found by integrating the normal curve equation. It would be extremely tedious to carry out an integration every time you wished to use the normal curve model. Fortunately, all necessary integrations have been performed and summarized in a convenient table, which shows the percent frequency of values within any distance (in terms of standard deviations) from the population mean. All that you are asked to do in order to use this table is to convert your raw scores to z values.† (Otherwise, an infinite number of tables would

---

* In addition to these two uses of the normal curve, many problems in inferential statistics are difficult to solve if the "technically correct" theoretical distribution is used. Often, however, the normal curve is a good approximation of the true distribution and can be used in its place, making matters simpler with little practical loss in accuracy.

† In the discussion of standard scores in Chapter 6, a capital Z was used to emphasize the fact that Z units do *not* change the shape of the original distribution and that the use of Z units does *not* require that the X distribution be normal. When the Z unit is used in conjunction with the normal curve, a lower-case z will be used.

be needed, one for each mean and standard deviation that a researcher might encounter.) This useful table is presented as Table B in the Appendix, and part of it is reproduced below.

	Percents Calculated from Area Under the Normal Curve					
z	.00	.01	.02	.03	. . .	.09
0.0						
⋮						
1.0	34.13	34.38	34.61	34.85		
1.1	36.43	36.65	36.86	37.08		
1.2	38.49	38.69	38.88	39.07		
1.3	40.32	40.49	40.66	40.82		
⋮						
∞						

The vertical column on the left represents z values expressed to one decimal place, while the top horizontal column gives the second decimal place. The values within the table represent the *percent area between the mean and the z value*, expressed to two decimal places. For example, to find the percent of cases between the mean and 1.03 standard deviations from the mean, select the 1.0 row and go across to the .03 column; the answer is read as 34.85%.

The table gives percents on one side of the mean only, so that the maximum percent given in the table is 50.00 (representing half of the area). However, the normal curve is *symmetric*, which implies that the height of the curve at any positive z value (such as $+1$) is exactly the same as the height at the corresponding negative value (-1). Therefore, the percent between the mean and a z of $+1$ is exactly the same as the percent between the mean and a z of -1, and presentation of a second half of the table to deal with negative scores is unnecessary. It is up to you to remember whether you are working in the half of the curve above the mean (positive z) or the half of the curve below the mean (negative z). Also, do not expect the value obtained from the table to be the answer to your statistical problem in every instance. Since the table only provides values between the mean and a given z score, some additional calculations will be necessary whenever you are interested in the area between the two z scores (neither of which falls at the mean) or the area between one end of the curve and a z score. These calculations will be illustrated below.

In Figure 9.2, the normal curve is presented with the X axis marked off to illustrate specified z distances from the mean. At the mean itself, the z value is 0 since the distance between the mean and itself is 0. The percent areas between the mean and other z values have been determined from Table B. Notice that the percent of the total area between μ (the population mean where $z = 0$) and a z of $+1.00$ is 34.13%, and the area between μ and $z = -1.00$ is also 34.13%. Thus, 68.26% of the total area under the

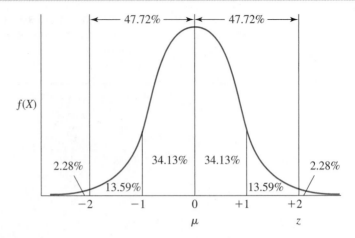

FIGURE 9.2
Normal curve: percent areas from the mean to specified z distances

normal curve lies between $z = -1.00$ and $z = +1.00$. Ninety-five percent of the normal curve lies within approximately two standard deviation units of μ (more exactly, between $z = -1.96$ and $z = +1.96$). So *if* the assumption of the normal curve model is correct, approximately 95% of all outcomes can be expected to fall between the mean and ± 2 standard deviations. For example, if the heights of American adult males were assumed to be normally distributed with the mean at 5 ft. 8 in. and the standard deviation equal to 3 in., you would expect close to 95% of these males to have heights between 5 ft. 8 in. $\pm$ 6 in., or between 5 ft. 2 in. and 6 ft. 2 in.

Characteristics of the Normal Curve

As shown in Figure 9.2, the shape of the normal curve looks like a bell; it is high in the middle and low at both ends or *tails.* Values in the middle of the curve (close to the mean of the distribution) have a greater probability of occurring than values further away (nearer the tails of the distribution). For example, you should expect z values between 0 and 1 to occur more frequently than z values between 1 and 2. Even though the score distances between the mean and $z = +1$ and between $z = +1$ and $z = +2$ are the same, the *area* between the mean and $z = +1$ is much greater than the area between $z = +1$ and $z = +2$.

As was observed in Chapter 2, the normal curve is unimodal (has one peak) and symmetric. However, not all curves that are bell shaped, unimodal, and symmetric are normal—that is, defined by the normal curve equation.

Notice that the normal curve does not touch the X axis until it reaches (+ and −) infinity. Extreme positive and negative values can occur, but are extremely unlikely. You *could* obtain z values of 5, 6, or more, but such z values would be very rare under this model (and might therefore suggest the possibility of an error in computation).

Illustrative Examples

The College Boards are administered each year to many thousands of high school seniors, and the scores are transformed in such a way as to yield a mean of 500 and a standard deviation of 100 (the same as SAT scores in Chapter 6). These scores are close to being normally distributed, so we can use the normal curve model in our calculations. Since we know both μ and σ, we can convert any College Board score (X) to a z score and refer it to Table B in the Appendix, using

$$z = \frac{X - \mu}{\sigma}$$

$$= \frac{X - 500}{100}$$

(In situations where μ and σ are not known, they can be estimated from the sample $\overline{X}$ and s provided that the sample size is large.)

There are many problems that can be solved by the use of this model:

1. What percent of high school seniors in the population can be expected to have College Board scores between 500 and 675?

The first step in any normal curve problem is to draw a rough diagram and indicate the answer called for by the problem; this will prevent careless errors. The desired percent is expressed as the shaded area in Figure 9.3.

FIGURE 9.3

Percent of population with College Board scores between 500 and 675

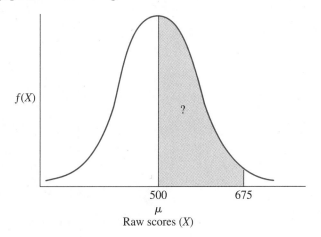

$f(X)$

?

500
μ
Raw scores (X)

675

To obtain the percent area between the mean (500) and a raw score of 675, convert the raw score to a z score:

$$z = \frac{675 - 500}{100} = +1.75$$

Table B reveals that the percent area between the mean and a z score of +1.75 is equal to 45.99%. Therefore, approximately 46% of the population is expected to have scores between 500 and 675. Thanks to the normal curve model, you can reach this conclusion even though you cannot measure the entire population.

2. What percent of the population can be expected to have scores between 450 and 500?

The diagram for this problem is shown in Figure 9.4. Converting the raw score of 450 to a z score yields:

$$z = \frac{450 - 500}{100} = -0.50$$

Table B does not include any negative z values, but entering the table at the z score of +.50 will produce the right answer because of the symmetry of the normal curve. The solution is that 19.15%, or about 19% of the population, is expected to have scores between 450 and 500.

3. What percent of the population can be expected to have scores between 367 and 540?

FIGURE 9.4

Percent of population with College Board scores between 450 and 500

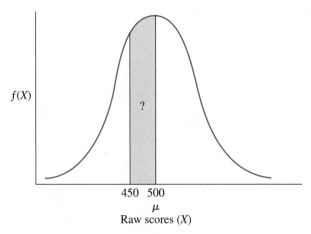

$f(X)$

?

450 500
μ
Raw scores (X)

This problem is likely to produce some confusion unless you use the diagram in Figure 9.5 as a guide. Table B gives values only between the mean and a given z score, so two steps are required. First obtain the percent between 367 and the mean (500); then obtain the percent between the mean and 540. Then *add* the two percents together to get the desired area. The z values are:

$$z = \frac{367 - 500}{100} = -1.33; \quad z = \frac{540 - 500}{100} + 0.40$$

Table B indicates that 40.82% of the total area falls between the mean and a z of -1.33, and 15.54% falls between the mean and a z of $+0.40$. So the percent of the population expected to have scores between 367 and 540 is equal to 40.82% + 15.54% or 56.36%.

4. What percent of the population can be expected to have scores between 633 and 700?

Once again, you are likely to encounter some difficulty unless you draw a diagram such as the one shown in Figure 9.6. The z score corresponding to a raw score of 633 is equal to $(633 - 500)/100$ or $+1.33$, and the z score corresponding to a raw score of 700 is equal to $(700 - 500)/100$ or $+2.00$. According to Table B, 40.82% of the curve falls between the mean and $z = +1.33$, and 47.72% of the curve falls between the mean and $z = +2.00$. So the answer is equal to 47.72% *minus* 40.82%, or 6.90%. In this example, subtraction (rather than addition) is the route to the correct answer, and drawing the illustrative diagram will help you decide on the proper procedure in each case.

FIGURE 9.5

Percent of population with College Board scores between 367 and 540

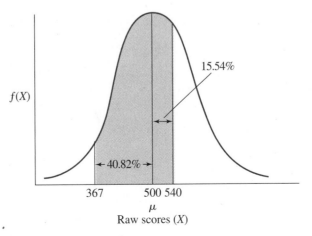

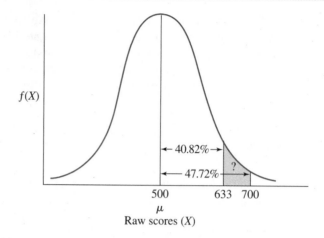

FIGURE 9.6

Percent of population with College Board scores between 633 and 700

5. What percent of the population can be expected to have scores above 725?

The percent to be calculated is shown in Figure 9.7. The z score corresponding to a raw score of 725 is equal to $(725 - 500)/100$ or $+2.25$, and the area between the mean and this z score as obtained from Table B is equal to 48.78%. At this point, avoid the incorrect conclusion that the value from the table is the answer to the problem. As the diagram shows, you need to find the percent *above* 725. Since the half of the curve to the right of the mean is equal to 50.00%, the answer is equal to 50.00% − 48.78%, or

FIGURE 9.7

Percent of population with College Board scores above 725

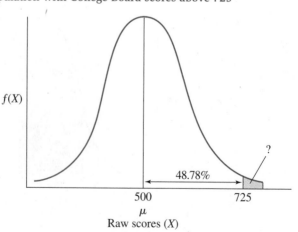

1.22%. A score of 725 is at about the 99th percentile, since about 1% of the population exceeds it.

6. What is the probability that a person drawn at random from the population will have a score of 725 or more?

This problem is identical to the preceding one, but is expressed in different terminology. As before, convert 725 to a z score, obtain the percent between the mean and 725, and subtract from 50% to obtain the answer of 1.22%. To express this as a probability, move the decimal point two places to the left. The probability that any one individual drawn at random from the population will have a score of 725 or more is .0122, or about .01.

7. A snobbish individual wishes to invite only the top 10% of the population on the College Boards to join a club that he is forming. What cutting score should he use to accept and reject candidates?

This problem is the reverse of the ones we have looked at up to this point. Previously, you were given a score and asked to find a percent; this time, the percent is specified and a score value is needed. The problem is illustrated in Figure 9.8.

The needed value is the raw score value corresponding to the cutting line in Figure 9.8. The steps are exactly the reverse of the previous procedure:

(1) If 10% are above the cutting line, then 40% (50% − 10%) are between the mean and the cutting line.

(2) Enter Table B in the *body* of the table (where the percents are given) with the value 40.00 (=40%). Read out the z score corresponding to the percent nearest in value to 40.00 (z = 1.28 for 39.97).

FIGURE 9.8

Raw score that demarcates the top 10% of the population on College Board scores

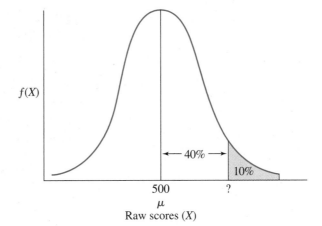

$f(X)$

← 40% →

10%

500

μ

Raw scores (X)

(3) Determine the *sign* of the z score. Since the diagram shows that the desired cutting score is *above* the mean, the z score is positive and equals + 1.28.

(4) Convert the z score to a raw score. Since $z = (X - \mu)/\sigma$, some simple algebraic manipulations yield $X = z\sigma + \mu$. Thus, $X = (+1.28)(100) + 500$, or 628. Any individual with a score of 628 or more may enter the elite club; individuals with scores of 627 or less are rejected.

8. What cutting score separates the top 60% of the population from the bottom 40%?

Since you are given a percent and asked to find a score, this is another "reverse" problem; you need to find the raw score illustrated in Figure 9.9.

The percent with which to enter the table is not immediately obvious, and a hasty reliance on any percent that seems to be handy (such as 60% or 40%) will produce an incorrect answer. The area between the mean and the cutting line is 10%, and it is this value that must be used to enter Table B. As in the previous problem, enter the table in the center; find the value nearest to 10.00 (here, 09.87), and read out the corresponding z score (0.25). Next, determine the sign. Since the diagram shows that the desired cutting score is *below* the mean, the sign is negative and $z = -0.25$. Finally, convert this value to a raw score: $X = (-0.25)(100) + (500)$, or 475. Thus the cutting score between the top 60% and the bottom 40% of the population is 475.

As you can see, the normal curve is useful for solving various types of problems involving scores and percentage distributions. Of even greater importance is its use with certain sampling distributions, such as means, to which we now turn.

FIGURE 9.9

Raw score demarcating the top 60% of the population on College Board scores

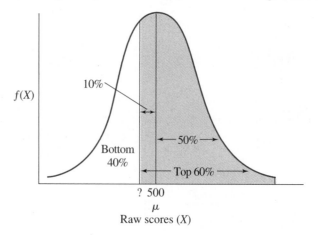

Summary

One frequently used theoretical model is the *normal curve.* One use of this model is as follows: If it is assumed that raw scores in the population are normally distributed, and that we know or can estimate the population mean and standard deviation, we can obtain (1) the percent of the population with raw scores above or below or between specified values and/or (2) the raw score that demarcates specified percents in the population.

1. *Type 1 problem: Given a raw score, find the corresponding frequency or percent frequency.*

    1. Convert the raw score to a z score: $z = (X - \mu)/\sigma$.
    2. Enter the normal curve table in the z column, and read out the percent area from the mean to this z value.
    3. Compute the answer to the problem. Note that the normal curve table gives only the area from the mean to the z value; be sure your final answer is the one called for by the problem.

2. *Type 2 problem: Given a frequency or percent frequency, find the corresponding raw score.*

    1. Compute the percent frequency (if it is not given) and enter the normal curve table in the area portion. Read out the corresponding z score. If the desired raw score is *below* the mean, assign a negative sign to this z score.
    2. Convert the z score to a raw score: $X = z\sigma + \mu$.

Chapter 10

Inferences About the Mean of a Single Population

PREVIEW

The Standard Error of the Mean

We wish to draw inferences about the mean of one population, based on a statistic whose sampling distribution is normal—namely, the sample mean. Why should the standard deviation of the raw scores *not* be used as the measure of variability? How does the variability of a distribution of *sample means* differ from that of *raw scores?*

What is the standard error of the mean? Why can't it be measured directly? How is it estimated?

What does the standard error of the mean tell us about the trustworthiness of a single sample mean as an estimate of the population mean?

What is sampling error?

Hypothesis Testing

What two hypotheses are made prior to conducting the statistical analysis?

Which of these hypotheses is assumed to be true?

What is a Type I error? A Type II error? Why can't we make the probability of both types of error as small as possible?

What consequences are likely to result from a Type I error? From a Type II error?

What is the power of a statistical test?

What is the criterion of significance, and how is it related to a Type I error?

What is the probability of a Type I error? Can the probability of a Type II error be stated as conveniently? What does this imply about how confident we can be when we retain the null hypothesis? When we reject the null hypothesis?

What is a two-tailed test of significance?

What advantage does the .05 criterion of significance have over the .01 criterion of significance? What is the advantage of the .01 criterion? Which is used more often in behavioral science research?

If the results of an experiment fall just short of the criterion of significance, what should be done?

(continued next page)

In the preceding chapter, a crucial assumption was made—that the *raw scores* in the population were normally distributed. This assumption enabled you to draw inferences about the percent of cases in the population between specified raw score values, even though you could not measure the entire population.

The present chapter deals with a second, more important use of the normal curve model: drawing inferences based on a *statistic* whose *sampling distribution* is normal. While the sampling distributions of several statistics are of normal form, we will use only the distribution of sample *means* to illustrate the use of the normal curve model. Our objective will be to draw inferences about the mean of a population (μ). For other statistics, the inferential strategy remains the same, but the details of the procedures are different.

The Standard Error of the Mean

Suppose that you are interested in studies of heights of American men, a variable known to be approximately normally distributed in the population. Also suppose that many, many random samples of the same size are selected from this population, and the mean of each sample is computed. The *sample means* when plotted as a frequency polygon will be normally distributed about *their* mean, which is also the population mean. Even if the distribution of raw scores in the population is *not* normal, the sample means will still be approximately normally distributed *if* the number of cases in each sample is sufficiently large. Thus the normal curve model can be used to represent the sampling distribution of the statistic $\overline{X}$.

Before you can proceed to draw inferences about the population mean, however, you need a measure of the variability of this sampling distribution. The standard deviation of the raw scores, σ (or its estimate s), is *not* the correct measure to use. To see why, consider once again the normal distribution of College Board scores in the population ($\mu = 500$). Selecting one person at random with a score of 725 is fairly unlikely. But drawing *a whole sample of people* with scores so extreme that the sample averages out to a mean of 725 is much less likely, since a greater number of low probability events must occur simultaneously. This implies that the variability of the distribution of sample means is *smaller* than the variability of the distribution of raw scores. In fact, it decreases as the sample N increases.

If a large number of sample means were actually available, you could measure the variability of the means directly. You would simply calculate the standard deviation *of the sample means,* using the usual formula for computing the standard deviation (Chapter 5) and treating the sample means just like ordinary numbers. However, behavioral scientists rarely have *many* samples of a given size. They usually have only one.

Fortunately, as a result of some mathematical-statistical theory, the variability of the distribution of sample means can be *estimated* from a *single* random sample drawn from the population. It can be shown that the *estimated* standard deviation of the sampling distribution of means of samples of size N, called the *standard error of the mean* and symbolized by $s_{\bar{x}}$, is given by:

$$s_{\bar{x}} = \frac{s}{\sqrt{N}}$$

where

s = standard deviation (based on the population variance
estimate) of the sample raw scores
N = number of observations in the sample

This quantity estimates the variability of means for samples of the given N, so it is a *standard deviation* of *means.* But since it does that, it also tells you how trustworthy is the single mean that you have, hence the name *standard error of the mean.* A *small* value of $s_{\bar{x}}$ indicates that if you were to draw many different random samples of the same size from this population, the various values of $\overline{X}$ would be relatively close to one another (since they would have a small standard deviation). Therefore, the single value of $\overline{X}$ that you have is likely to be a rather accurate estimate of μ. But if $s_{\bar{x}}$ is *large,* and you were to draw many different random samples of the same size from this population, the various values of $\overline{X}$ would differ markedly from one another (since they would have a large standard deviation). Therefore, any one value of $\overline{X}$—including the single value that you have—is a less trustworthy indication of the true value of μ.

As you can see from the formula, the standard error of the mean must be smaller than the standard deviation of the raw scores. It becomes smaller as the sample size grows larger. The term *standard error* signifies that the difference between a population mean and the mean of a sample drawn from that population is an "error," caused by the cases that happened by chance to be included in the sample. If *no* error existed (as would happen if the "sample" consisted of the entire population and $N = \infty$), all of the sample means would exactly equal each other, and the value of the standard error of the mean would be zero. This situation would be ideal for purposes of statistical inference, since the mean of any one sample would exactly equal the population mean. Sad to say, it never happens; actual data are always subject to sampling error.

To clarify the preceding discussion, let us assume that the mean height in the population of American men is equal to 68 inches, and compare the distribution of raw scores to the distribution of sample means. A hypothetical distribution of five million height observations for American men is shown in Figure 10.1; each X represents one *raw score.* Now suppose 1,000 random samples of $N = 100$ observations each are randomly selected from

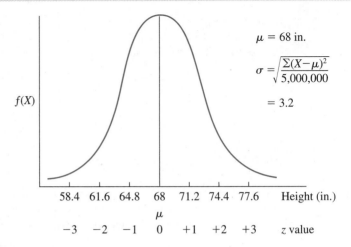

$\mu = 68$ in.

$$\sigma = \sqrt{\frac{\Sigma(X-\mu)^2}{5{,}000{,}000}}$$

$f(X)$

$= 3.2$

58.4	61.6	64.8	68	71.2	74.4	77.6	Height (in.)

μ

-3	-2	-1	0	$+1$	$+2$	$+3$	z value

FIGURE 10.1

Distribution of observations of heights (in inches) for the population of 5,000,000 adult males

$X_1 = 68.3$ in.	$X_2 = 66.2$ in.	$X_3 = 73.1$ in.	. . .	$X_{5{,}000{,}000} = 69.3$ in.
Observation 1	*Observation 2*	*Observation 3*		*Observation 5,000,000*
or	*or*	*or*		*or*
Individual 1	*Individual 2*	*Individual 3*		*Individual 5,000,000*

the population, and the mean of each sample is calculated. The resulting distribution of *sample means* is illustrated in Figure 10.2.

Notice that the shape of *both* distributions (original distribution of raw scores or single observations, and distribution of sample means) is normal and centered about μ (68 in.). However, the distribution of sample means is *less variable* than is the distribution of raw scores. The standard deviation of the raw scores shown in Figure 10.1 would be calculated directly by applying the usual formula for the standard deviation to the 5,000,000 raw scores; let us suppose that σ is found to be equal to 3.2 in. The standard deviation of the sample means in Figure 10.2 would be calculated directly by applying the usual formula for the standard deviation to the 1,000 sample means. As the diagram indicates, this is equal to only 0.32 in.* The means of the samples, each based on $N = 100$ observations, will have a σ equal to one-tenth $(=1/\sqrt{N})$ the σ of the individual observations.

Although this highly unrealistic situation is useful for purposes of illustrating the standard error of the mean, let us return to the real world where you have only one sample at your disposal. Suppose that you randomly

---

* When the population standard deviation is known, the standard error of the mean is symbolized by $\sigma_{\bar{x}}$ and is found from $\sigma_{\bar{x}} = \sigma/\sqrt{N}$, where σ = the standard deviation of the population raw scores and N = the number of observations in the random sample. Thus, in the present case, the value of $\sigma_{\bar{x}}$ would be $3.2/\sqrt{100}$ or .32.

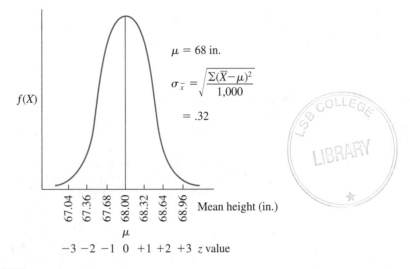

$\mu = 68$ in.

$$\sigma_{\bar{x}} = \sqrt{\frac{\Sigma(\bar{X}-\mu)^2}{1,000}}$$

$= .32$

$f(X)$

67.04 67.36 67.68 68.00 68.32 68.64 68.96   Mean height (in.)

μ

-3 -2 -1 0 $+1$ $+2$ $+3$ z value

FIGURE 10.2

Distribution of mean heights (in inches) for 1,000 randomly selected samples of $N = 100$ height observations in each sample

$\bar{X}_1 = 68.1$ in. $\bar{X}_2 = 67.8$ in. $\bar{X}_3 = 68.4$ in.   . . . $\bar{X}_{1,000} = 67.7$ in.
 Sample 1   Sample 2   Sample 3   Sample 1,000

draw *one* sample of $N = 100$ *midwestern* American men, and you calculate the mean height of this sample as $\bar{X} = 67.4$ in. and the standard deviation of the raw scores as $s = 3.5$ in. Then, the standard error of the mean can be *estimated* as follows:

$$s_{\bar{x}} = \frac{s}{\sqrt{N}} = \frac{3.5}{\sqrt{100}} = .35$$

You would like to know whether the average height of midwestern men is different from the average of the entire U.S. adult male population (68 in.). Without measuring the entire population of midwestern men, you cannot answer this question with certainty. However, you can use the sample mean and the standard error of the mean to draw an inference about the population. There are two possibilities: *either* the sample with a mean of 67.4 comes from a population with a mean of 68, and the difference of .6 between these two values is due to sampling error (that is, the cases that happened to fall in the sample of 100 midwestern men that you selected at random had a mean that fell .6 below the population mean) *or* the sample does *not* come from a population with a mean of 68, but from one whose mean is some other value. In order to decide between these alternatives, it would be helpful to know how *unusual* such a sample would be — that is, the probability of selecting a sample with $\overline{X} = 67.4$ *if* $\mu = 68$. Which of the two alternatives should be selected, based on the experimental data?

Hypothesis Testing

One way to answer this question is to *test the hypothesis* that the mean of the population from which the sample of 100 midwestern men was drawn is equal to 68 inches. The testing of a statistical hypothesis is designed to help make a decision about a population parameter (here, μ).

In Chapter 8, we saw that if a series of experiments is run wherein a coin is flipped six times, six heads out of six flips should be obtained approximately once or twice in 100 experiments *if* the coin is fair (that is, under the hypothesis $P(H) = 1/2$). Since this was "sufficiently unlikely" in terms of the ".05 decision rule," the hypothesis that the coin was fair (Hypothesis 0) was *rejected* in favor of the hypothesis that the coin was biased (Hypothesis 1). The procedure is similar in the present height study: it is necessary to determine the probability of drawing a sample with a mean of 67.4 inches *if* the population mean is in fact 68 inches, and to compare this probability to a decision rule.

The Null and Alternative Hypotheses

The first step is to state the *null hypothesis* (symbolized by H_0), which specifies the hypothesized population parameter. In the present study,

$$H_0: \mu = 68$$

This null hypothesis implies that the sample, with mean equal to 67.4 inches, is a random sample from the population with μ equal to 68 inches (and that the difference between 67.4 and 68 is due to sampling error). As in the coin experiment, the probability of obtaining a sample mean of the observed value (67.4) is calculated under the assumption that H_0 is true (that is, that $\mu = 68$).

Next, an *alternative hypothesis* (symbolized by H_1) is formed. It specifies another value or set of values for the population parameter. In the present study,

$$H_1: \mu \neq 68$$

This alternative hypothesis states that the population from which the sample comes does *not* have a μ equal to 68 inches. That is, the difference of .6 between the sample mean and the null-hypothesized population mean is due to the fact that the null hypothesis is false.

Consequences of Possible Decisions

If the probability of obtaining a sample mean of 67.4 when $\mu = 68$ is deemed "sufficiently unlikely" to occur by a decision rule such as the ".05 rule," H_0 is rejected in favor of H_1. If this probability is so large as to make H_0 "not sufficiently unlikely," H_0 is retained. While the null is *hypothesized* to be true at the outset, the statistical test *never proves* whether the null hypothesis is true or false. The test merely indicates whether or not H_0 is "sufficiently unlikely" given the decision rule. Without measuring the entire population, it is impossible to *prove* anything, and the final decision that you reach could be correct or incorrect. In fact, there are four possible outcomes:*

1. The population mean is actually 68 (that is, the sample comes from a population where $\mu = 68$) and you *incorrectly reject* H_0. Rejecting a *true* null hypothesis is called a *Type I Error,* and the probability (or *risk*) of committing such an error is symbolized by the Greek letter alpha (α).

2. The population mean is actually *not* 68 and you *incorrectly retain* H_0. Failing to reject a *false* null hypothesis is called a *Type II Error,* and the probability (risk) of committing this error is symbolized by the Greek letter beta (β).

3. The population mean is actually *not* 68 and you *correctly reject* H_0. The probability of reaching this correct decision is called the *power* of the statistical test, and is equal to $1 - \beta$.

4. The population mean is actually 68 and you *correctly retain* H_0. The probability of making this correct decision is equal to $1 - \alpha$.

The four eventualities are summarized in Table 10.1.

Whenever you conduct a statistical test, there are always two possible kinds of error risk: rejecting a null hypothesis that is true (alpha risk), and

TABLE 10.1

Model for error risks in hypothesis testing

		State of the Population	
		H_0 Is Actually True	H_0 Is Actually False
outcome of experiment dictates:	*retain H_0*	correct decision: probability of retaining true H_0 is $1 - \alpha$	type II error: probability (risk) of retaining false H_0 is β
	reject H_0	type I error: probability (risk) of rejecting true H_0 is α	correct decision: probability of rejecting false H_0 (power) is $1 - \beta$

---

* Compare this discussion with that of flipping coins with a friend in Chapter 8.

failing to reject a null hypothesis that is false (beta risk). An important aspect of such statistical tests is that it is possible to specify and control the probability of making each of the two kinds of errors. For the present, we will be concerned primarily with Type I error and the corresponding alpha risk. Type II error and beta risk will be treated in Chapter 14.

Why can't these errors be summarily dismissed with an instruction to make the probability of their occurrence as small as possible? For a sample of a given size, making a Type I error *less* likely simultaneously makes a Type II error *more* likely, and vice versa. Therefore, you must balance the desirable aspects of reducing the probability of one kind of error against the undesirable aspects of increasing the probability of the other kind of error.

This is a problem of real practical concern to the behavioral scientist. The null hypothesis gets its name from the idea that "nothing is going on." That is, the null hypothesis states that there is *no* effect or *no* difference of the kind that the experiment is seeking to establish. Usually, the theory that the scientist hopes to support is identified with the *alternative* hypothesis. Rejecting H_0 will cause the scientist to conclude that his theory has been supported, and failing to reject H_0 will cause him to conclude that his theory has not been supported. This is a sensible procedure, since scientific caution dictates that the researcher should not jump to conclusions. He should claim success only when there is a strong indication to that effect—that is, when the results are sufficiently striking to cause rejection of H_0.

If the researcher commits a Type I error (rejects H_0 when it is in fact true), he will triumphantly conclude that his theory has been supported when the theory is incorrect. This will cause him professional embarrassment when other researchers attempt to replicate his findings, but are unable to do so. Therefore, the researcher might be tempted to make the probability of a Type I error as small as possible. But this will increase the chances of a Type II error, which also has rather drastic consequences. If the researcher fails to reject H_0 when it is false, he will fail to conclude that his theory is valid. He will miss out on the chance to advance the state of scientific knowledge, and he will fail to receive the recognition due him for his correct theory. This problem is resolved in part by the "decision rule" that is selected.

The Criterion of Significance

Having set up the null and alternative hypotheses, we need to define *how unusual* a sample mean must be to cause you to reject H_0 and accept H_1. Intuitively, an observed sample mean of 68.1 would hardly represent persuasive evidence against the null hypothesis that $\mu = 68$, even though it is not exactly equal to the hypothesized population value. Conversely, an observed sample mean of 40 would do more to suggest that the hypothesized population value was incorrect. Where should the line be drawn? As we pointed out in Chapter 8, there are no *absolute* rules for establishing the values for which H_0 *must* be rejected. All that the statistical test can do is to compare hypotheses about a population parameter, and yield a statement about the probability of an experimental result *if H_0 is true.*

The numerical value specified by the decision rule that you select is called the *criterion of significance.* When you choose the ".05 decision rule," you are using the 5% or *.05 criterion of significance.* That is, you define an observed probability of .05 or less as *sufficiently* unlikely to reject H_0. Since the criterion of significance is denoted by α, the .05 criterion can be expressed in symbols as $\alpha = .05$.

As was indicated previously, the value of α specifies the probability of making a Type I error. When you use the .05 criterion of significance, the probability of falsely concluding that $\mu \neq 68$ is equal to .05. This means that if you were to repeat this experiment many times when the null hypothesis is true, an average of one out of every 20 decisions would be to (incorrectly) reject the null hypothesis when you use the .05 criterion. (Note that the probability of falsely retaining the null hypothesis that $\mu = 68$, a Type II error, is not so conveniently determined. See Chapter 14.)

You should reject H_0 if the experimental result is unusually deviant (high or low). This is called a *"two-tailed" test of the null hypothesis* using the .05 criterion of significance; it is illustrated in Figure 10.3. Note that the circumstances portrayed are for the null hypothesis to be *true,* and the probability of falsely concluding that $\mu \neq 68$ is equal to $2\frac{1}{2}\% + 2\frac{1}{2}\% = .05$. When the normal curve model is appropriate, the cutoff scores beyond which H_0 should be rejected are easily ascertained using the normal curve table (Table B). Since 2.5% of the area of the curve is in each tail, $50.0 - 2.5\%$ or 47.5% falls between the mean and the cutoff score. Entering Table B in the center with a value of 47.5%, the cutting scores expressed as z values are found to be equal to -1.96 and $+1.96$. Once the sample mean has been expressed as

FIGURE 10.3

Areas of rejecting H_0 in both tails of the normal sampling distribution using the .05 criterion of significance when H_0 actually is true

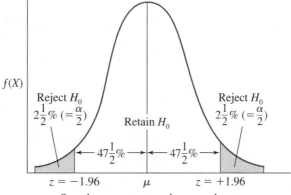

a z value, the .05 criterion of significance states that H_0 should be rejected if z is less than or equal to -1.96 or z is greater than or equal to $+1.96$. Conversely, H_0 should be retained if z is between -1.96 and $+1.96$. Note that the area in each rejection region is equal to $\alpha/2$, or .05/2 or .025, and that the total *area of rejection* (or *critical region*) is equal to the significance criterion α or .05.

Another traditional, but equally arbitrary, criterion is the 1% or *.01 criterion of significance.* Using this criterion, two-tailed hypotheses about the population mean are rejected if, when H_0 is true, a sample mean is so unlikely to occur that no more than 1% of sample means would be so extreme. Here again, $\alpha/2$ or .005 (or 1/2 of 1%) is in each tail. This criterion is illustrated in Figure 10.4.

When the normal curve model is appropriate, the cutoff scores beyond which H_0 should be rejected using the .01 criterion are ascertained by entering Table B in the body of the table with the value 49.5%. These cutting scores, expressed as z values, are -2.58 and $+2.58$. Once the sample mean has been expressed as a z value, the .01 criterion of significance states that H_0 should be rejected if z is less than or equal to -2.58 or z is greater than or equal to $+2.58$. Conversely, H_0 should be retained if z is between -2.58 and $+2.58$. Once again, the area of rejection is equal to α (0.5% + 0.5% = 1% or .01), the criterion of significance.

The 1% criterion of significance is a more stringent test for rejecting H_0 than is the 5% criterion, since only *very* unusual sample means will be considered "sufficiently unlikely." It has the corresponding advantage of making the probability of a Type I error smaller. However, the 1% criterion is more lenient about retaining H_0, so the probability of a Type II error is greater. When you use the 1% criterion, you will reject H_0 only in more extreme sit-

FIGURE 10.4
Areas of rejecting H_0 in both tails of the normal sampling distribution using the .01 criterion of significance when H_0 is true

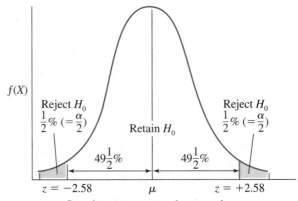

Sample mean expressed as a z value

uations (when $P \leq .01$), so you are more likely to be correct (and avoid a Type I error) when you do reject H_0. But you must therefore fail to reject H_0 in some rather unlikely situations (that is, whenever $P > .01$, even if it is less than .05), so you are more likely to be wrong when you do retain H_0. (Still more stringent criteria of significance, such as the .001 criterion, would further reduce the chances of a Type I error at the cost of further increasing the chances of a Type II error.)

The most frequently used criterion of significance in behavioral research is the .05 criterion, which keeps the chances of a Type II error (missing out on a new finding) lower than for the .01 criterion. It is generally felt that this criterion is sufficiently conservative in terms of avoiding a Type I error. (Although a .10 criterion of significance would reduce the chances of a Type II error, it is usually regarded as too likely to lead to a Type I error to be acceptable.) The value of .05 is a highly arbitrary choice, however, so don't be rigid about values that are very close to the cutting line. If for example $z = 1.88$, the decision rule requires you to retain H_0. It would be a shocking violation of sound research practice (and ethics) to slide the cutting score down a few notches in order to reject H_0 and support your theory, but it is frustrating to fall just short of the critical value of $+1.96$. The sensible thing to do in such a close case is to repeat the experiment if possible, so as to get more data on which to base a decision. (This problem may be minimized by using procedures discussed in Chapter 14.)

The Statistical Test for the Mean of a Single Population When σ Is Known

Let us suppose that you select the .05 criterion of significance in the height experiment. Also assume that you know that the *population* standard deviation, σ, is equal to 3.2. (In behavioral science research, this assumption is rarely justified. We will discuss the procedures that are used when σ is *not* known in the next section.) Then, the sample mean must be expressed as a z value in order to compare it to the criterion values of ± 1.96. However, since you are drawing inferences about the *mean* of a single population, the *standard error of the mean* must be used as the measure of variability. That is,

$$z = \frac{\overline{X} - \mu}{\sigma_{\overline{x}}}$$

where

$\overline{X} =$ the observed value of the sample mean
$\mu =$ the hypothesized value of the population mean
$\sigma_{\overline{x}} =$ the standard error of the mean when σ is known $(= \sigma/\sqrt{N})$

Given that $\sigma = 3.2$,

$$\sigma_{\bar{x}} = \frac{3.2}{\sqrt{100}} = .32$$

$$z = \frac{67.4 - 68}{.32}$$

$$= \frac{-.60}{.32}$$

$$= -1.875$$

The z value of -1.875 falls between the critical values of -1.96 and $+1.96$, and is *not* in the area of rejection. Therefore, your decision should be to retain H_0.

Since the probability of a Type II error (failing to reject H_0 when H_0 is actually false) is not conveniently determined, you do not know how confident to be in your decision to retain H_0. You should conclude only that *there is not sufficient reason to reject the null hypothesis* that the average height for midwestern American men is equal to the average height for American men in general. (The determination of the probability of a Type II error is discussed in Chapter 14.) In other words, a sample mean of 67.4 is "not sufficiently unlikely" to occur if $\mu = 68$ for you to reject H_0. This result is said to be *"not statistically significant at the .05 level."*

Now suppose instead that it is known that $\sigma = 3.0$. In this situation,

$$\sigma_{\bar{x}} = \frac{3.0}{\sqrt{100}} = .30$$

$$z = \frac{67.4 - 68}{.30}$$

$$= -2.00$$

This z value of -2.00 falls beyond the critical value of -1.96 and *is* in the area of rejection. So here you should reject H_0 in favor of H_1.

In this instance, you should conclude that the average height for midwestern American men is *not* equal to (in fact, is less than) the average height for American men in general. The statistical test has indicated that a sample mean of 67.4 is "sufficiently (less than .05) unlikely" to occur if $\mu = 68$ for you to reject H_0. That is, if H_0 were true and many, many random samples of size 100 were drawn from the population, you would expect to get a sample mean that deviated from 68 by 0.6 (or more) less than 5% of the time. Consequently, it is a better bet to abandon the assumption that H_0 is true (and that $\mu = 68$) and prefer the alternative hypothesis.

This result is said to be *"statistically significant at the .05 level of confidence."* In other words, H_0 is rejected at the .05 significance level.

The Statistical Test for the Mean of a Single Population When σ Is Not Known: The *t* Distributions

To use the procedure discussed above, you must know the value of the *population* standard deviation. Only rarely, however, is σ known in behavioral science research. Most of the time, it is necessary to compute the estimated standard error of the mean based on the *estimated* standard deviation (s), where $s_{\bar{x}} = s/\sqrt{N}$.

Although this formula may look similar to the preceding one, there is an important difference. Values obtained from the formula $(\bar{X} - \mu)/s_{\bar{x}}$ are *not* normally distributed (although the normal curve will yield a good approximation if the sample size is greater than 25 or 30). This is because $s_{\bar{x}}$ is only an estimate of $\sigma_{\bar{x}}$. So using the normal curve procedure will give wrong answers, primarily when the sample size is small, because the statistical model is incorrect. You must instead refer to the exact distribution of $(\bar{X} - \mu)/s_{\bar{x}}$, which is appropriate for all sample sizes but particularly necessary for small samples. This new statistical model is known as the *t distributions*.

In contrast to the normal curve, there is a *different t* distribution for every sample size. The sampling distribution of means based on a sample size of 10 has one *t* distribution, while the sampling distribution of means based on a sample size of 15 has a different *t* distribution. Fortunately, you do not have to know the shape of each of the various *t* distributions. Probability tables for the curves corresponding to each sample size have been determined by mathematicians. All you need do is note the size of your sample, and then refer to the proper *t* distribution. Such a *t* table is presented as Table C in the Appendix. (The assumptions of *t*-distributions are discussed in Chapter 11.)

Degrees of Freedom

The *t* table is given in terms of *degrees of freedom (df)* rather than sample size. When you have a sample mean and a hypothesized population mean (and σ is not known):

$$df = N - 1$$

That is, degrees of freedom in the present situation are equal to *one less* than the sample size. Recall that this is equal to the divisor of s^2 (Chapter 5).

"Degrees of freedom" is a concept that arises at several points in statistical inference; you will encounter it again in connection with chi square (Chapter 17). The number of degrees of freedom is the number of freely varying quantities in the kind of repeated random sampling that produces sampling distributions. In connection with the test of the mean of a single population, it refers to the fact that each time a sample is drawn and the population variance is estimated, it is based on only $N - 1$ df. That is, there are $N - 1$ freely varying quantities. When you find for each of the N observations its deviation from the sample mean, $X - \overline{X}$, not all of these N quantities are free to vary in any sample. Since they must sum to zero (as was shown in Chapter 4), the Nth value is automatically determined once any $N - 1$ of them are fixed at given observed values. For example, if the sum of $N - 1$ deviations from the mean is -3, the Nth must equal $+3$ so that the sum of all deviations from the mean will equal zero. So in this situation, there are only $N - 1$ df, and therefore the t distribution is here based on $N - 1$ df. The df and not the N is the fundamental consideration, because df need not be $N - 1$ in other applications of t.

The Statistical Test

A section of the t table is shown below:

df	. . .	$t_{.05}$	$t_{.01}$	. . .
⋮				
4		2.78	4.60	
⋮				
10		2.23	3.17	
11		2.20	3.11	
12		2.18	3.06	
13		2.16	3.01	
14		2.15	2.98	
15		2.13	2.95	
⋮				
40		2.02	2.70	
⋮				
120		1.98	2.62	
⋮				
∞		1.96	2.58	

Each horizontal row represents a separate distribution that corresponds to the df. The row for ∞ represents the normal distribution.

To illustrate the use of the t table, suppose that you wish to test the null hypothesis H_0: $\mu = 17$ against the alternative H_1: $\mu \neq 17$ using the .05 criterion of significance. You obtain a sample of $N = 15$ and find that $\overline{X} = 18.80$ and $s = 3.50$. Then compute

$$t = \frac{\overline{X} - \mu}{s_{\overline{x}}}$$

where

$\overline{X}$ = the observed value of the sample mean
μ = the hypothesized value of the population mean
$s_{\overline{x}}$ = the estimated standard error of the mean ($= s/\sqrt{N}$)

In the present example,

$$t = \frac{18.80 - 17}{(3.50/\sqrt{15})} = \frac{1.80}{.90} = 2.00$$

You now compare the *t* value that you have computed to the critical value obtained from the *t* table. For 14 degrees of freedom (one less than your *N* of 15) and using the two-tailed .05 criterion of significance, the critical value is 2.15. A *t* value larger than this *in absolute value* (that is, ignoring the sign) is "sufficiently unlikely" to occur if H_0 is true for you to reject H_0. That is, the probability of obtaining *t* values absolutely larger than 2.15 when $df = 14$ is .05. Therefore, if the absolute value of the *t* that you compute is *less than* 2.15, *retain H_0*. If, however, your result is *greater* in absolute value than 2.15, *reject H_0* in favor of H_1. (In the unlikely event that your computed *t* value exactly equals the *t* value obtained from the table, reject H_0.) In our example, the obtained *t* value of 2.00 is not larger in absolute value than 2.15. So you should retain H_0, and conclude that the data do *not* warrant concluding that the population mean is not 17.

If instead your sample size were 12, you would have 11 degrees of freedom. The corresponding critical value from the table, again using the .05 criterion of significance, would be 2.20. You would retain H_0 if the absolute size of the *t* value you computed were less than 2.20, and you would reject H_0 in favor of H_1 if this value were equal to or greater in absolute value than 2.20. Similarly, for the same sample size, the critical value of *t* using the .01 criterion of significance is 3.11. Here, retain H_0 if your computed *t* value is no greater than 3.11 in absolute value and reject H_0 otherwise.

The *t* and *z* Distributions Compared

The *t* distribution for nine degrees of freedom and the normal curve are compared in Figure 10.5. Note the fatter tails in the case of the *t* distribution, so that you have to go further out on the tails to find points that set off the .05 or .01 areas of rejecting H_0. While 95% of the normal curve lies between the *z* values -1.96 and $+1.96$, 95% of the *t* distribution for nine degrees of freedom lies between -2.26 and $+2.26$. So using the two-tailed .05 criterion, 2.26 (and *not* 1.96) must be used as the critical value when σ is not known and $df = 9$.

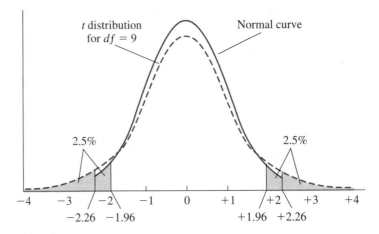

A comparison between the normal curve and the t distribution for $df = 9$

Notice also in the t table that the *smaller* the sample size, the *larger t* must be in order to reject H_0 at the same criterion of significance. Also, when the sample size is large, z and t are virtually equivalent. Thus, for 40 degrees of freedom, $t = 2.02$ (using the .05 criterion). For 120 degrees of freedom, $t = 1.98$. These values are so close to the z value of 1.96 that it makes little difference which one you use. Suppose that in the study of heights of midwestern American men discussed earlier in this chapter, a sample of 100 midwestern men is found to have $\overline{X} = 67.4$ inches and $s = 3.5$ inches and σ is *not* known. To test the null hypothesis that $\mu = 68$, you would compute

$$s_{\overline{x}} = \frac{3.5}{\sqrt{100}} = .35$$

$$t = \frac{67.4 - 68}{.35}$$

$$= -1.71$$

You would decide to retain H_0 whether you compared 1.71 (the absolute value) to the critical t value of 1.99 for $df = 99$, or to the z value of 1.96. In behavioral science research, it is customary to report the obtained value as a t value even for large samples. The reason is that although z is a good approximation in large samples, it is still an approximation. So you should use the critical t value obtained from the t table, rather than the z value of 1.96 for the .05 criterion or 2.58 for the .01 criterion, unless you happen to know the exact value of σ. *Retain H_0* if your computed t value (ignoring the sign) is smaller than the critical value, and *reject H_0* otherwise.

Interval Estimation

Hypothesis testing is very popular in the behavioral sciences. However, this method has several drawbacks. One difficulty is that when the results are statistically significant, the researcher is entitled to reject *only* the exact value specified by the null hypothesis.

Suppose that a researcher wishes to test the hypothesis that students at a particular high school differ from the average IQ (namely, a score of 100). In this study, the null hypothesis states that these students come from a population where $\mu = 100$. The alternative hypothesis states that $\mu \neq 100$. The researcher obtains a sample of 50 students, finds that the mean IQ for this sample is 115.72, performs the appropriate *t* test, and finds that the results are statistically significant using the .05 criterion of significance. The researcher therefore rejects the null hypothesis, switches to the alternative hypothesis, and concludes that the mean IQ of all students in this high school is greater than 100.

Although the researcher was able to reject the null hypothesis that $\mu = 100$, the researcher *cannot* reject the hypothesis that $\mu = 102$, that $\mu = 101$, or even that $\mu = 100.1$. The research study tested one and only one null hypothesis, namely $\mu = 100$, so only this hypothesis can be rejected when the results are statistically significant. To be sure, it is likely that values very close to the null-hypothesized value of 100 should also be rejected. But it is not certain, nor is it clear what "very close" means numerically.

If the population mean does happen to be 101 or 102, the students in this high school are actually about average in intelligence. A person with an IQ of 101 or 102 does not behave very differently from a person with an IQ of 100; a difference of one or two IQ points is virtually meaningless. Despite the statistically significant results, important issues remain unresolved. The students in this sample are very unlikely to have come from a population where $\mu = 100$. But it may be likely that they have come from a population where μ is so close to 100 that, for all practical purposes, it is the same. Without additional information, there is no way to tell.

One good way to resolve this difficulty is by estimating a *confidence interval* within which the population mean is likely to fall. Rather than specifying a single numerical value (as does H_0), a confidence interval specifies a *range of values* within which a parameter (here, the population mean) is likely to fall. How likely? If you are using the .05 criterion of significance, you can be 95% confident that the population mean falls within the range specified by the confidence interval. The end points of the confidence interval are called *confidence limits*.

Computation

To illustrate the use of confidence intervals, consider the height experiment discussed previously in this chapter. We wish to determine an interval that will include all "not sufficiently unlikely" values of μ (all values for which

H_0 should be retained). Using the .05 criterion of significance and $df = 99$, the furthest that any sample mean can lie from μ and still be retained as "not sufficiently unlikely" is $\pm 1.99 s_{\bar{x}}$ (where 1.99 is the critical value of t for the specified df, 99). So the lower limit of the confidence interval may be found by solving the following equation for μ:

$$+1.99 = \frac{\overline{X} - \mu}{s_{\bar{x}}}$$

The upper limit of the confidence interval may be found by solving the following equation for μ:

$$-1.99 = \frac{\overline{X} - \mu}{s_{\bar{x}}}$$

This is readily done as follows:

$$\pm 1.99 s_{\bar{x}} = \overline{X} - \mu$$
$$\mu = \overline{X} \pm 1.99 s_{\bar{x}}$$

or

$$\overline{X} - 1.99 s_{\bar{x}} \leq \mu \leq \overline{X} + 1.99 s_{\bar{x}}$$

Thus,

$$67.4 - 1.99(.35) \leq \mu \leq 67.4 + 1.99(.35)$$
$$67.4 - .70 \leq \mu \leq 67.4 + .70$$
$$66.7 \leq \mu \leq 68.1$$

You should retain any hypothesized value of μ between 66.7 inches and 68.1 inches, inclusive, using the confidence interval corresponding to the .05 criterion of significance (called the *95% confidence interval*). Conversely, reject any hypothesized value of μ outside this interval.

The general formula for computing confidence intervals when σ is not known is:

$$\overline{X} - ts_{\bar{x}} \leq \mu \leq \overline{X} + ts_{\bar{x}}$$

where

t = critical value from the t table for $df = N - 1$
$\overline{X}$ = observed value of the sample mean

The confidence interval corresponding to the .01 criterion of significance is called the *99% confidence interval*. In the height experiment, the critical value of t for $df = 99$ and the .01 criterion is 2.62. Thus, the 99% confidence interval is equal to

$$67.4 - 2.62(.35) \leqslant \mu \leqslant 67.4 + 2.62(.35)$$

$$67.4 - .92 \leqslant \mu \leqslant 67.4 + .92$$

$$66.48 \leqslant \mu \leqslant 68.32$$

Note that the 99% confidence interval is *wider* than the 95% confidence interval. You can be more sure that the population mean falls in the 99% confidence interval, but you must pay a price for this added confidence: the interval is larger, so μ is less precisely estimated. (You could be absolutely certain that the population mean falls within the "confidence interval" $-\infty \leqslant \mu \leqslant +\infty$, but such a statement is useless—it tells nothing that you did not know beforehand.) For $\alpha = .05$, you can be 95% confident that the true population mean falls within the stated interval (narrower, hence more informative). For $\alpha = .01$, you can be 99% confident (more sure) that the true population mean falls within the stated interval (wider, hence less informative).

If σ *is known,* use the critical z values (1.96 for the 95% confidence interval, 2.58 for the 99% confidence interval) instead of t.

Confidence Intervals and Null Hypothesis Tests

The literature dealing with statistical inference in the behavioral sciences, and the scientific journals, place great emphasis on testing null hypotheses. Nevertheless, interval estimation has important advantages.

Confidence intervals allow you to do everything that null-hypothesis testing does. If a value of μ specified by a null hypothesis falls within the confidence interval, that null hypothesis is retained. If a value of μ specified by a null hypothesis falls outside the confidence interval, that null hypothesis is rejected. And confidence intervals do more, because they provide a range of population means that should be retained. In the IQ experiment, for example, a 95% confidence interval of 110.62–120.82 would indicate that the students in this high school are clearly above average in intelligence. Here, the smallest hypothesized value of μ that should be retained is 110.62. If instead the 95% confidence interval is 101.60–129.84, the researcher will need to be more cautious. While rather high values of μ are reasonably likely, so too are values of μ that are very close to 100—so close as to represent virtually average intelligence.

Confidence intervals provide considerably more information than the results of testing a specific null hypothesis. Therefore, confidence intervals should be used more widely in behavioral science research.

The Standard Error of a Proportion

Suppose that a few weeks before an election, a worried politician takes a poll by drawing a random sample of 400 registered voters. He finds that 53% intend to vote for him, while 47% prefer his opponent.* He is pleased to observe that he has the support of more than half of the sample, but he knows that the sample may not be an accurate indicator of the population because an unrepresentative number of his supporters may have been included by chance. What should he conclude about his prospects in the election?

This problem is similar to the preceding ones in this chapter. There is one population in which the politician is interested (registered voters), and he wishes to draw an inference about the mean of this population based on data obtained from a sample. In particular, he would like to know if the percent supporting him is greater than 50% (in which case he will win the election). There is one important difference in the present situation, however: his data are in terms of percents or *proportions*.

While the general strategy is the same as in the case of the standard error of the mean, special techniques must be used to deal with proportions. The needed formula is

$$z = \frac{p - \pi}{\sqrt{\pi(1 - \pi)/N}}$$

where

p = **proportion observed in the sample**
π = **hypothesized value of the** *population* **proportion**
N = **number of people in the sample**

The denominator of this formula, $\sqrt{\pi(1 - \pi)/N}$, is the *standard error of a proportion*, symbolized by σ_p. It serves a similar purpose to the standard error of the mean, but it is a measure of the variability of *proportions* in samples drawn at random from a population where the proportion in question is π. (Notice that in this formula, π is used to represent the hypothesized value of the population proportion. It does *not* represent the usual mathematical value of 3.14159.)

Since the population π is specified, and the σ of the population in question is known (it can be proved equal to $\sqrt{\pi(1 - \pi)}$), the results are referred to the z table. In the case of the anxious politician, the critical population value in which he is interested is .50: he wants to know whether the

---

* For simplicity, we assume that there are no undecided or "won't say" voters in this sample. There will be such in reality, and their omission makes the possibly incorrect assumption that they will break the same way as those who make a choice.

observed sample proportion of .53 is sufficiently different from .50 to enable him to conclude that a majority of the population of voters will vote for him. Thus,

$$\pi = \text{hypothesized population proportion} = .50$$

$$p = \text{proportion observed in sample} = .53$$

The statistical analysis is as follows:

$$H_0: \pi = .50$$

$$H_1: \pi \neq .50$$

$$\alpha = .05$$

$$z = \frac{.53 - .50}{\sqrt{.50(1 - .50)/400}} = \frac{.03}{.025} = 1.20$$

The value of 1.20 is smaller than the critical z value of 1.96 needed to reject H_0. Therefore, the politician *cannot* conclude that he will win the election. It is "not sufficiently unlikely" for 53% of a sample of 400 voters to support him, even if only 50% of the population will vote for him. The politician will therefore have a nervous few weeks until the votes are in, and he may well conclude from the results that he should increase his campaigning efforts. The results are illustrated in Figure 10.6.

There is one important caution regarding the use of the standard error of the proportion. When the hypothesized value of the population proportion (π) is quite large or small and the sample size is small, computing a z value

FIGURE 10.6

Sampling distribution of p when $\pi = .50$ and $N = 400$

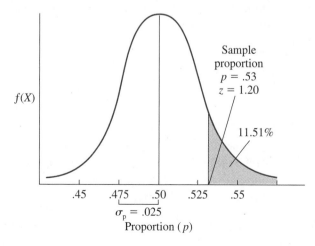

and referring it to the normal curve table will yield incorrect results. As a rough guide, you should *not* use this procedure when either $N\pi$ or $N(1 - \pi)$ is less than 5. For example, if you wish to test the null hypothesis that $\pi = .85$ with a sample size of 20,

$$N\pi = (20)(.85) = 17 \quad \text{and} \quad N(1 - \pi) = (20)(.15) = 3$$

Since $N(1 - \pi)$ is less than 5, the procedures discussed in this section should *not* be used. The tested value of π is too extreme, and the sample size is too small, to give an accurate answer. You should either use the appropriate statistical techniques to deal with this situation (the binomial distribution, discussed in more advanced texts) or increase the size of your sample.

Confidence Intervals for π

Suppose you wish to determine all reasonably likely values of the population proportion (π) using the .05 criterion of significance. The confidence interval is

$$(p - 1.96\sigma_p) \leqslant \pi \leqslant (p + 1.96\sigma_p)$$

There is one difficulty: σ_p depends on the value of π, which is not known. In the previous section, a value of π was specified by H_0 and used in the calculation of σ_p. However, there is no single hypothesized value of π in interval estimation.

One possibility, which yields a fairly good approximation *if the sample size is large*, is to use p as an estimate of π. With a large sample, p is unlikely to be so far from π as to greatly affect σ_p and hence the size of the confidence interval. In the case of the nervous politician in the previous section, $p = .53$; and $\sqrt{.53(1 - .53)/N}$ is equal to .025. Then, the 95% confidence interval is equal to

$$.53 - 1.96(.025) \leqslant \pi \leqslant .53 + 1.96(.025)$$

$$.53 - .049 \leqslant \pi \leqslant .53 + .049$$

$$.48 \leqslant \pi \leqslant .58$$

Since the sample size is large ($N = 400$), the politician can state with 95% confidence that between approximately 48% and 58% of the population will vote for him. Note that the value of .50 *is* within the confidence interval, as would be expected from the fact that the null hypothesis that $\pi = .50$ was retained. To determine the 99% confidence interval, the procedure would be the same except the z value of 2.58 would be used instead of 1.96.

"Exact" solutions for the confidence interval for π (which do not require using p as an estimate of π and hence large samples) do exist, but involve techniques that are beyond the scope of this book.

One-Tailed Tests of Significance

The statistical tests discussed earlier in this chapter (see for example Figures 10.3, 10.4, and 10.5) are called *two-tailed* tests of significance. This is true for both confidence intervals and null-hypothesis tests. This means that any hypothesis regarding the population mean, μ, is rejected if it falls beyond either the lower *or* the upper limit of the confidence interval. Or a specific H_0 is rejected if a z or t value is obtained that is either extremely high (far up in the upper tail of the curve) *or* extremely low (far down in the lower tail of the curve). Consequently, there is a rejection area equal to $\alpha/2$ in each tail of the distribution. As we have seen, the corresponding null and alternative hypotheses are:

H_0: $\mu =$ a specific value

H_1: $\mu \neq$ this value (that is, μ is greater than the value specified by H_0 *or* is less than this value)

Let us suppose that a researcher argues as follows: "My theory predicts that the mean of population X is *less than* 100." (For example, she may predict that children from homes in which there are few books and little emphasis on verbal skills are below average on a standardized test of intelligence.) "It's all the same to me whether μ_x is equal to 100 or much greater than 100; my theory is disconfirmed in either case. If I use a two-tailed test, I am forced to devote $2\frac{1}{2}\%$ of my rejection region (using $\alpha = .05$) to an outcome that is meaningless insofar as my theory is concerned. Instead, I will place the entire 5% rejection region in the lower tail of the curve." (See Figure 10.7.) "In other words, I will test the following null and alternative hypotheses:

$$H_0: \mu_x \geqslant 100$$
$$H_1: \mu_x < 100$$
$$\alpha = .05$$

"Note that I am being properly conservative by assuming at the outset that my results are due to chance (the null hypothesis). Only if I obtain a result very unlikely to be true if H_0 is true will I conclude that my theory is supported (switch to the alternative hypothesis). Looking at the t table for $df = 500$ (since I have $N = 501$ and σ is not known), I see that a t value of -1.65 places 5% of the distribution in the lower tail. Therefore, I will retain

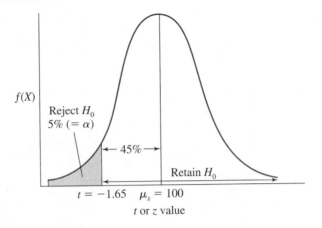

FIGURE 10.7

One-tailed test of the null hypothesis that $\mu_x \geq 100$ against the alternative that $\mu_x < 100$

H_0 if t is anywhere between -1.65 and $+\infty$ and reject H_0 if t is -1.65 or less. Since a two-tailed test of significance would lead to rejection of H_0 only if t were -1.96 or less (or $+1.96$ or more), the one-tailed value of -1.65 quite properly makes it easier for me to reject H_0 (and conclude that my theory is supported) when the results are in the direction that I have predicted."

Similar reasoning could be applied if the researcher claimed to be interested *only* in the outcome that the population mean was *greater* than 100. The entire rejection region would be placed at the upper end of the curve, and H_0 would be retained if t were between $-\infty$ and $+1.65$ and rejected if t were $+1.65$ or more.

One good way to resolve the issue of whether to use one-tailed tests is by establishing a confidence interval. In this example, the researcher would set up a 95% confidence interval by using the t value that places 5% of the distribution in the lower tail, namely 1.65. The reason for using this value, rather than the usual t value of 1.97 for a 95% confidence interval and $df = 500$, is that the researcher is focusing entirely on one direction: if her theory is correct, the population mean is *less than* 100. As we observed previously in this chapter, using the smaller t value of 1.65 produces a narrower confidence interval than would the t value of 1.97, which enables the researcher to estimate the value of μ more precisely.

To illustrate, let us consider four possible confidence intervals that this researcher might obtain:

(1) $82.3 \leq \mu \leq 90.4$

(2) $92.7 \leq \mu \leq 100.8$

(3) $97.2 \leqslant \mu \leqslant 105.3$

(4) $106.4 \leqslant \mu \leqslant 114.5$

The first set of results supports the researcher's theory. She can conclude with 95% confidence that the population mean is well below 100—namely, between 82.3 and 90.4. While it is unlikely that the population mean is less than 82, it is also unlikely that the mean is close to 100.

The second set of results is less encouraging. To be sure, the population mean is likely to be less than 100. But the researcher cannot reject the hypothesis that μ is very close to 100 or even slightly higher, since the confidence interval extends as far as 100.8. Because there is little difference in behavior between an IQ of 98 or 99 and an IQ of 100, this confidence interval includes quite a few values that do not provide much support for the researcher's theory. Yet because most of the confidence interval is below 100, she may be on the right track. One reasonable course is to design new experiments that deal with this issue in a somewhat different way.

The third set of results is discouraging. It is unlikely that the population mean is much below 100. Therefore, these results provide very little support for the researcher's theory.

The fourth set of results poses a major problem. Since 100 falls below the lower confidence limit, it may seem as though we can conclude that the population mean is likely to be *greater* than 100. However, these results are in the *opposite* direction to that predicted by the researcher. Such opposite results may well have important implications. But they cast serious doubt on the researcher's decision to use the one-tailed test that she chose, *and this confidence interval is based on that decision* (because the t value of 1.65 was used). Yet the design of an experiment cannot be changed after the results are in. It is too late to switch to a t value of 1.97, compute a new confidence interval, and see if it also exceeds the value of 100.

Note that the researcher would have had the same problem had she conducted a null-hypothesis test. She would have chosen the same t value of 1.65, because she placed the entire rejection region in the lower tail of the curve, and it would be too late to change this decision once the results were in.

In such situations, we recommend that the experiment be repeated with a new sample of subjects and the t value that corresponds to a two-tailed test (e.g., 1.97 for $df = 500$) before any conclusions are drawn. This is the "price" that the researcher must pay for her incorrect prediction.

Summary

1. Hypothesis testing—general considerations

1. State the *null hypothesis* (denoted by the symbol H_0) and the *alternative hypothesis* (denoted by H_1). It is important to understand

that you cannot *prove* whether H_0 or H_1 is true because you cannot measure the entire population.

2. Begin with the assumption that H_0 is true. Then obtain the data and test out this assumption. If this assumption is unlikely to be true, you will abandon it and "switch your bets" to H_1; otherwise, you will retain it.

Before you collect any data, it is necessary to define in numerical terms what is meant by "unlikely to be true." In statistical terminology, this is called selecting a *criterion* (or *level*) *of significance*, represented by the symbol α.

a. Using the .05 criterion of significance, "unlikely" is defined as having a probability of .05 or less. In symbols, $\alpha = .05$.

b. Using the .01 criterion of significance, "unlikely" is defined as having a probability of .01 or less. In symbols, $\alpha = .01$.

Other criteria of significance can be used, but these are the most common.

3. Having selected a criterion of significance, obtain your data and compute the appropriate statistical test. If the test shows that H_0 is unlikely to be true, reject H_0 in favor of H_1. Otherwise, retain H_0.

4. Though you are backing the hypothesis indicated by the statistical test (H_0 or H_1), you could be wrong. You could make either of the following two kinds of error, depending on your decision:

a. Type I Error: Reject H_0 when H_0 is in fact true. This error is less likely when the .01 criterion of significance is used. In fact, the probability of this kind of error (that is, the risk you run that your decision to reject H_0 is incorrect) is equal to the chosen criterion of significance α.

b. Type II Error: Retain H_0 when H_0 is in fact false. This error is less likely when the .05 criterion of significance is chosen. The probability of this kind of error, denoted by the symbol β, is not so conveniently determined. (See Chapter 14.)

Thus, the relative importance of each kind of error helps to determine what criterion of significance to use.

2. The standard error of the mean

To test null hypotheses about the *mean of one population,* first compute the estimated *standard error of the mean.* This is a measure of how accurate the sample mean is likely to be as an estimate of the population mean. Then, use the standard error of the mean to determine whether the difference between the observed sample mean and the hypothesized value of the population mean is or is not "sufficiently unlikely" to occur if H_0 is true (that is, if the hypothesized value of the population mean is correct).

The formula for estimating the standard error of the mean from a single sample is

$$s_{\bar{x}} = \frac{s}{\sqrt{N}}$$

where

$s_{\bar{x}}$ = **the estimated standard error of the mean**
s = **standard deviation of the sample of observations**

Then, to draw inferences about the population mean, use

$$t = \frac{\overline{X} - \mu}{s_{\bar{x}}} \qquad \text{with} \qquad df = N - 1$$

Use z instead of t if σ is known. For large samples, t and z are approximately equal.

3. Confidence intervals

Confidence intervals establish all reasonably likely values of a population parameter, such as a mean or proportion. Confidence intervals provide considerably more information than the results of testing a specific null hypothesis, because they show the decision that should be made for every conceivable null-hypothetical value rather than only one.

Use $s_{\bar{x}}$ when determining confidence intervals for the mean of one population. Obtain the critical value of t from the t table ($df = N - 1$; $\alpha = 1.0 -$ confidence) and compute

$$(\overline{X} - ts_{\bar{x}}) \leq \mu \leq (\overline{X} + ts_{\bar{x}})$$

where

t = **critical value from t table**

Use z instead of t if σ is known. For large samples, z and t are approximately equal.

4. The standard error of a proportion

If the data are in the form of proportions,

1. Compute

$$\sigma_p = \sqrt{\frac{\pi(1 - \pi)}{N}}$$

2. Compute

$$z = \frac{p - \pi}{\sigma_p}$$

where

σ_p = the standard error of a proportion
p = proportion observed in the sample
π = hypothesized value of the population proportion

Do *not* use this procedure if $N\pi$ or $N(1 - \pi)$ is less than 5.

5. One-tailed tests of significance

The use of one-tailed tests of significance is sometimes justified in behavioral science research. If the results are in the direction predicted by the researcher, it is more likely that statistical significance will be obtained. But if the results are in the opposite direction, the experiment must be repeated using a two-tailed design before any conclusions are drawn. Here again, the use of confidence intervals is recommended.

Chapter 11

Testing Hypotheses About the Difference Between the Means of Two Populations

PREVIEW

The Standard Error of the Difference

We wish to draw inferences about the *difference* between the means of *two* populations. What is the correct standard error term to use in this situation? Why can't it be measured directly?

What is an experimental group? A control group?

Why is it desirable to use random samples?

How does sampling error affect the difference between the means of the experimental group and the control group?

What does the standard error of the mean tell us about the trustworthiness of a single difference between two sample means as an estimate of the difference between the population means?

Estimating the Standard Error of the Difference

What are the procedures for estimating the value of the standard error of the difference?

How is the standard error of the difference related to the variance of the population? To the size of the sample?

The *t* Test for Two Sample Means

What is the correct statistical model to use in this situation?

How do we test hypotheses about the difference between the means of two populations (independent samples)?

Measures of the Strength of the Relationship Between the Two Variables

When we obtain a statistically significant value of *t* for the difference between two means, why is it desirable to convert this value to a measure that shows the strength of the relationship between the two variables?

Confidence Intervals for $\mu_1 - \mu_2$

How do we estimate an interval in which the difference between the two population means is likely to fall?

(continued next page)

PREVIEW (continued)

What is gained by using confidence intervals?

Using the *t* Test for Two Sample Means: Some General Considerations

How confident can we be if we decide to retain the null hypothesis?
How confident can we be if we decide to reject the null hypothesis?
What assumptions underlie the use of this *t* test?

The *t* Test for Matched Samples

How do matched samples differ from independent samples? When is it desirable to use matched samples?
How do we test hypotheses about the difference between the means of two matched populations?

Summary

In the preceding chapter, we looked at techniques for drawing inferences about the mean of one population. More often, scientists wish to draw inferences about differences between two or more populations. For example, a social psychologist may want to know if the mean of the population of urban Americans on a pencil-and-paper test of spatial abilities is equal to the mean of the population of suburban Americans. Or, an experimental psychologist may wish to find out whether rats perform better on a discrimination learning task to gain a reward or to avoid punishment—that is, whether the mean number of correct responses is greater for the "reward" population or for the "punishment" population. In both cases, the question of interest concerns a comparison between the means of *two* samples.

In this chapter, we will discuss techniques for drawing inferences about the *difference between the means of two populations* $(\mu_1 - \mu_2)$. Procedures for drawing inferences about the means of more than two populations, or about other aspects of two or more populations, will be considered in later chapters.

The Standard Error of the Difference

Suppose that you wish to conduct a research study to determine whether or not the use of caffeine improves performance on a college mathematics examination. To test this hypothesis, you obtain two *random* samples of college students taking mathematics, an *experimental group* and a *control group.* (Random sampling is one good way to avoid groups that are greatly different in ability, motivation, or any other variable that might obscure the effects of the caffeine.) You then give all members of the experimental group a small dose of caffeine one hour prior to the test. At the same time, the control group is given a placebo (a pill that, unknown to them, has no biochemical effect whatsoever). This is to control for the possibility that administering *any* pill will affect the students' performance. (For example, they may become more psychologically alert and obtain higher scores.) Thus, the control group serves as a baseline against which the performance of the experimental group can be evaluated. Finally, the mean test scores of the two groups are compared. Suppose that the mean of the experimental group equals 81, and the mean of the control group equals 78. Should you conclude that caffeine is effective in improving the test scores?

As was shown in Chapter 10, any sample mean is almost never exactly equal to the population mean because of sampling error—those cases that happened, by chance, to be included in the sample. Therefore, the two populations in question (a hypothetically infinite number of test-takers given caffeine and a hypothetically infinite number of test-takers given a placebo) may have equal means even though the sample means are different. Consequently, you cannot tell what conclusion to reach just by looking at the sample means. You need additional information: namely, whether the three-point difference between 81 and 78 is likely to be a trustworthy indication that the population means are different (and that caffeine *has* an effect), or whether this difference may be due solely to which cases happened to fall in each sample (in which case you should *not* conclude that caffeine has an effect).

How can this additional information be obtained? In Chapter 10 we saw that the variability of the sampling distribution of means drawn from one population, s_X, provides information as to how trustworthy is any one $\overline{X}$ as an estimate of μ. The present problem concerns how trustworthy is any one *difference*, $\overline{X}_1 - \overline{X}_2$, as an estimate of the difference between the population means, $\mu_1 - \mu_2$. This problem is also solved by making use of an appropriate sampling distribution: the sampling distribution of the *difference* between *two* sample means.

Suppose that a variable is normally distributed in each of two defined populations, and that the populations have equal means and equal variances (that is, $\mu_1 = \mu_2$ and $\sigma_1^2 = \sigma_2^2$). If many pairs of random samples of equal size were drawn from the two populations, a distribution of differences between the paired means could be established empirically. For each pair, you would subtract the mean of the second sample $(\overline{X}_2)$ from the mean of the first sample $(\overline{X}_1)$. Since the means of the two populations are equal, any difference between the sample means must be due solely to sampling error.

Figure 11.1 illustrates the procedure needed to obtain an empirical sampling distribution of 1,000 differences in the case where $\mu_1 = \mu_2 = 80$ and the size of each random sample is 30. The resulting frequency distribution is shown in Table 11.1, and the frequency polygon plotted from this

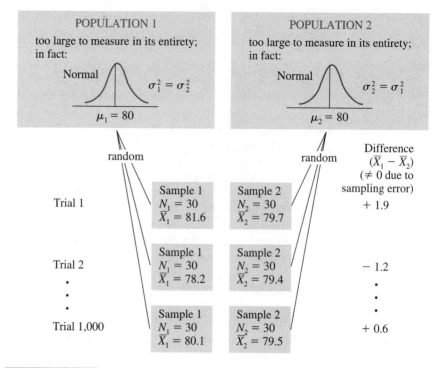

FIGURE 11.1

Illustration of procedure for obtaining the empirical sampling distribution of differences between two means ($N = 30$)

distribution is shown in Figure 11.2. The sample means are not in general exactly equal to 80, and the differences between pairs of means are not in general exactly equal to zero, due to sampling fluctuations. Notice that the distribution is approximately symmetric, which indicates that chance differences are equally likely to occur in either direction—that is, $(\bar{X}_1 - \bar{X}_2) > (\mu_1 - \mu_2)$ or $(\bar{X}_1 - \bar{X}_2) < (\mu_1 - \mu_2)$. Thus the mean of the distribution of differences tends to be equal to $\mu_1 - \mu_2$, which in this case is zero.

Given the data in Table 11.1, you could calculate the standard deviation of the distribution of differences for the given sample size by using the usual formula for the standard deviation (Chapter 5). This standard deviation would tell you how much, on the average, a given $\bar{X}_1 - \bar{X}_2$ is likely to differ from the "true" difference of zero (the central point of the distribution of differences). Consequently, it would indicate how trustworthy is the single difference that you have on hand as an estimate of $\mu_1 - \mu_2$. This standard deviation is therefore called the *standard error of the difference.*

A relatively *large* standard error of the difference indicates that any single difference between a pair of sample means must be viewed with grave sus-

TABLE 11.1

Empirical sampling distribution of 1,000 differences
between pairs of sample means ($N = 30$) drawn
from two populations where $\mu_1 = \mu_2 = 80$
(hypothetical data)

Difference Between Sample Means $\overline{X}_1 - \overline{X}_2$	Number of Samples (f)
Greater than +11.49	0
+10.50 to +11.49	1
+9.50 to +10.49	0
+8.50 to +9.49	1
+7.50 to +8.49	4
+6.50 to +7.49	7
+5.50 to +6.49	21
+4.50 to +5.49	32
+3.50 to +4.49	54
+2.50 to +3.49	77
+1.50 to +2.49	107
+.50 to +1.49	122
−.50 to +.49	153
−1.50 to −.51	114
−2.50 to −1.51	95
−3.50 to −2.51	82
−4.50 to −3.51	60
−5.50 to −4.51	31
−6.50 to −5.51	22
−7.50 to −6.51	10
−8.50 to −7.51	4
−9.50 to −8.51	1
−10.50 to −9.51	2
Smaller than −10.50	0
Total	1,000

picion. Since large discrepancies between $\overline{X}_1$ and $\overline{X}_2$ are likely *even if they come from populations with equal means,* you need a sizable difference before you can safely conclude that μ_1 is not equal to μ_2. If, on the other hand, the standard error of the difference is relatively *small,* you can place more confidence in any one sample difference as an estimate of the population difference. Large discrepancies between $\overline{X}_1$ and $\overline{X}_2$ are *not* likely to occur (and mislead you) if in fact $\mu_1 = \mu_2$, so a large difference between the sample means *is* a trustworthy sign that the population means differ.

In practice, behavioral scientists never have many paired random samples of a given size. Therefore, it is impossible to compute the standard error of the difference empirically. Instead, as in Chapter 10, you must make use of estimation procedures developed by statisticians. In the present situation, you need techniques that enable you to *estimate* the standard error of the difference based on the *one* pair of samples at your disposal.

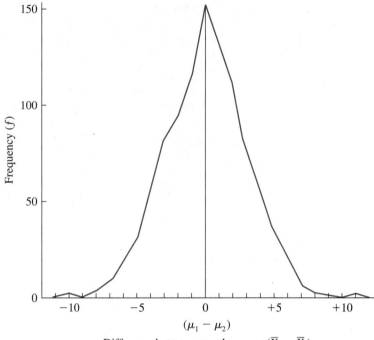

FIGURE 11.2

Frequency polygon of data in Table 11.1

Estimating the Standard Error of the Difference

As was the case with $s_{\bar{X}}$, developing the estimation formula for the standard error of the difference by means of a formal proof is beyond the scope of this book. Instead, we will attempt to explain the formula in nonmathematical terms.

Since it is assumed that the two population variances are equal, the variances of each of the two samples may be combined into a single estimate of the common value of σ^2. If the sizes of the two samples are exactly equal, you can obtain this combined or *pooled* estimate by computing the average of the two sample variances. Often, however, the sample sizes are not equal, in which case greater weight must be given to the larger-sized sample. That is, a *weighted* average must be computed. The general formula for the *pooled variance* (symbolized by s^2_{pooled}), which may be used for equal or unequal sample sizes, is:

$$s^2_{\text{pooled}} = \frac{(N_1 - 1)s_1^2 + (N_2 - 1)s_2^2}{N_1 + N_2 - 2}$$

where

$$s_1^2 = \text{population variance estimate of sample 1}$$
$$s_2^2 = \text{population variance estimate of sample 2}$$
$$N_1 = \text{number of cases in sample 1}$$
$$N_2 = \text{number of cases in sample 2}$$

Then, the estimation formula for the standard error of the difference is:

$$s_{\bar{X}_1 - \bar{X}_2} = \sqrt{\frac{s_{\text{pooled}}^2}{N_1} + \frac{s_{\text{pooled}}^2}{N_2}}$$

$$= \sqrt{s_{\text{pooled}}^2 \left(\frac{1}{N_1} + \frac{1}{N_2} \right)}$$

where

$$s_{\bar{X}_1 - \bar{X}_2} = \text{estimated standard error of the difference}$$

This formula should not look totally unfamiliar. We saw in Chapter 10 that the estimation for the standard error of the mean ($s_{\bar{X}}$) is equal to $s/\sqrt{N}$. Consequently, $s_{\bar{X}}^2$ is equal to s^2/N. Here, since a difference is subject to sampling error from *two* means, the sampling error variance reflects both sources of error. In addition, it uses a more stable estimate of the population variance by pooling the variance information from the two samples.

It should be obvious that the larger the variance of the population (as estimated by s_{pooled}^2), the larger the estimated standard error of the difference. Only slightly less obvious is the fact that the *larger* the sample size, the *smaller* the standard error. Differences based on large samples will have smaller sampling error (and are thus likely to be more accurate estimates of the population difference) than are differences based on small samples, just as is the case for a single mean.

The two steps given above can be combined into a single formula, as follows:

$$s_{\bar{X}_1 - \bar{X}_2} = \sqrt{\frac{(N_1 - 1)s_1^2 + (N_2 - 1)s_2^2}{N_1 + N_2 - 2} \left(\frac{1}{N_1} + \frac{1}{N_2} \right)}$$

The t Test for Two Sample Means

When you test hypotheses about the difference between two population means, the correct statistical model to use is the t distributions. Once the estimated standard error of the difference has been computed, a t value (and therefore a probability value) may be found for any given obtained difference, as follows:

$$t = \frac{(\overline{X}_1 - \overline{X}_2) - (\mu_1 - \mu_2)}{s_{\overline{X}_1 - \overline{X}_2}}$$

Notice that the two sample means yield a single difference score, $\overline{X}_1 - \overline{X}_2$, which is compared to the null-hypothesized mean of the difference scores, $\mu_1 - \mu_2$.

It is possible to test any hypothesized difference between the means of two populations. For example, you could test the hypothesis that the mean of the first population is 20 points greater than the mean of the second population by setting $\mu_1 - \mu_2$ equal to 20 in this equation. Much more often than not, however, you will want to test the hypothesis that the means of the two populations are equal (that is, $\mu_1 = \mu_2$ or $\mu_1 - \mu_2 = 0$). In this situation, the t formula can be simplified to

$$t = \frac{\overline{X}_1 - \overline{X}_2}{s_{\overline{X}_1 - \overline{X}_2}}$$

or

$$t = \frac{\overline{X}_1 - \overline{X}_2}{\sqrt{\dfrac{(N_1 - 1)s_1^2 + (N_2 - 1)s_2^2}{N_1 + N_2 - 2} \left(\dfrac{1}{N_1} + \dfrac{1}{N_2} \right)}}$$

$$df = N_1 + N_2 - 2$$

The degrees of freedom on which this t value is based, which is needed to obtain the critical value of t from the t table, is determined from the fact that the population σ^2 is estimated using $N_1 - 1$ degrees of freedom from sample 1 and $N_2 - 1$ degrees of freedom from sample 2 (see Chapter 10). Therefore, the estimate is based on the combined degrees of freedom:

$$(N_1 - 1) + (N_2 - 1) \quad \text{or} \quad N_1 + N_2 - 2$$

Returning to the problem of caffeine and mathematics test scores posed at the beginning of this chapter, the first step in the statistical analysis consists of stating the hypotheses and establishing a significance criterion, α:

$$H_0: \mu_{\text{exp}} = \mu_{\text{con}}$$

$$H_1: \mu_{\text{exp}} \neq \mu_{\text{con}}$$

$$\alpha = .05$$

H_0 states that the experimental and control samples come from populations with equal means, while H_1 states that these two samples come from populations with different means. The customary .05 criterion of significance is specified.

Next, let us suppose that the experimental results are as follows:

Experimental Group	Control Group
$N_1 = 22$	$N_2 = 20$
$\overline{X}_1 = 81.0$	$\overline{X}_2 = 78.0$
$s_1^2 = 12.0$	$s_2^2 = 10.0$

The estimated standard error of the difference is equal to:

$$s_{\overline{X}_1 - \overline{X}_2} = \sqrt{\frac{(21)(12) + (19)(10)}{22 + 20 - 2}\left(\frac{1}{22} + \frac{1}{20}\right)}$$

$$= \sqrt{(11.05)\left(\frac{1}{22} + \frac{1}{20}\right)}$$

$$= 1.03$$

The *t* value is equal to:

$$t = \frac{81.0 - 78.0}{1.03} = 2.91, \qquad df = 22 + 20 - 2 = 40$$

Referring to the *t* table for 40 degrees of freedom, we find that a *t* value greater in absolute value than 2.02 is needed to justify rejection of H_0 using the .05 criterion of significance. Since the obtained value of *t* is greater in absolute value than the critical value obtained from the table, it is *not* likely that the obtained difference between the sample means is due to sampling error. So you reject H_0, and conclude that the experimental and control groups are random samples from populations with *different* means. More specifically, caffeine *has* a positive effect on mathematics test scores.

The rejection areas for this experiment are illustrated in Figure 11.3.

FIGURE 11.3

Acceptance and rejection regions for caffeine experiment in the *t* distribution for $df = 40$

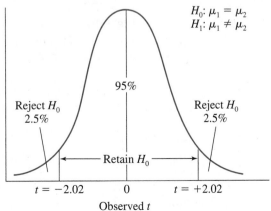

$H_0: \mu_1 = \mu_2$
$H_1: \mu_1 \neq \mu_2$

95%

Reject H_0
2.5%

Reject H_0
2.5%

Retain H_0

$t = -2.02$ 0 $t = +2.02$

Observed *t*

Measures of the Strength of the Relationship Between the Two Variables

When you obtain a statistically significant value of t for the difference between two means, it is very desirable to convert this value to an index that will show the *strength* of the relationship between the two variables. For example, the statistically significant value of t obtained in the caffeine experiment indicates that caffeine has a positive effect on mathematics test scores. It does *not* indicate whether the relationship between the presence or absence of caffeine and test scores is weak, moderate, or strong. Even a large t does not necessarily mean a strong effect.

To obtain this important information about the strength of the relationship between caffeine and test scores, it is necessary to convert the obtained value of t to a statistic called the point biserial correlation coefficient. Since we have not yet considered the fundamentals of correlation, we will defer further discussion of this topic until Chapter 13.

Confidence Intervals for $\mu_1 - \mu_2$

The use of confidence intervals to determine all reasonably likely values of the mean of a *single* population was discussed in Chapter 10. In a similar fashion, all reasonably likely values of the *difference between two population means* can be found by using the following confidence interval:

$$[(\overline{X}_1 - \overline{X}_2) - ts_{\overline{X}_1 - \overline{X}_2}] \leq \mu_1 - \mu_2 \leq [(\overline{X}_1 - \overline{X}_2) + ts_{\overline{X}_1 - \overline{X}_2}]$$

where

**t = critical value obtained from the t table for the specified
criterion of significance for the appropriate df**

Since this confidence interval pertains to the *difference* between two population means, the appropriate error term is the standard error of the difference $(s_{\overline{X}_1 - \overline{X}_2})$.

For example, in the caffeine experiment, the observed sample means were 81.0 for the experimental group and 78.0 for the control group; the standard error of the difference was 1.03; and the critical value of t for $df = 40$ and $\alpha = .05$ was 2.02. The 95% confidence interval for the difference between the population means is:

$$[(81.0 - 78.0) - (2.02)(1.03) \leq \mu_1 - \mu_2$$

$$\leq [(81.0 - 78.0) + (2.02)(1.03)]$$

$$(3.0 - 2.08) \leq \mu_1 - \mu_2 \leq (3.0 + 2.08)$$

$$.92 \leq \mu_1 - \mu_2 \leq 5.08$$

You can state with 95% confidence that the mean of the experimental population on the mathematics test is included in the interval that runs from

.92 points to 5.08 points greater than the mean of the control population. Note that zero is *not* in this interval, indicating that it is *not* likely that $\mu_1 = \mu_2$.

To determine the 99% confidence interval, the procedure would be the same except that the critical value of *t* for 40 *df* for $\alpha = .01$, which is 2.70, would be used instead of 2.02. The 99% confidence interval would therefore be:

$$(3.0 - 2.78) \leqslant \mu_1 - \mu_2 \leqslant (3.0 + 2.78)$$

$$.22 \leqslant \mu_1 - \mu_2 \leqslant 5.78$$

As always, the 99% confidence interval is wider than the 95% confidence interval. You can be more certain that the true difference between the two population means falls in the 99% confidence interval, but you must pay a price for this added confidence: the interval is larger, so $\mu_1 - \mu_2$ is estimated less precisely.

Here again, confidence intervals allow you to do everything that null-hypothesis testing does. If a value of $\mu_1 - \mu_2$ specified by the null hypothesis (such as the frequently used value of zero) falls within the confidence interval, that null hypothesis is retained. If a value of $\mu_1 - \mu_2$ specified by the null hypothesis falls outside the confidence interval, that null hypothesis is rejected. And by specifying a range of population differences that should be retained, confidence intervals provide us with more information.

For example, in the caffeine experiment, a 95% confidence interval of 7.62–13.74 would indicate that the effect of caffeine on mathematics test scores is a probable mean increase of at least seven points and not more than fourteen points. However, a 95% confidence interval of 0.62–20.74 would tell a different story. Here, caffeine may cause an increase of as much as 20 points or as little as one point. Since a one-point increase is virtually negligible, the possibility remains that caffeine has no important practical effect on mathematics test scores. (Such a wide confidence interval suggests the likelihood of substantial sampling error. One good solution is to repeat the experiment with a larger sample.) Yet in both cases, a null-hypothesis test will yield identical results: it will tell us only that the null-hypothesized value of zero between the two population means should be rejected.

Therefore, confidence intervals should be used more widely in behavioral science research dealing with the difference between two population means. When statistically significant values of *t* are obtained, important additional information can also be obtained by converting these values to measures that show the strength of the relationship between the two variables. (See Chapter 13.)

Using the *t* Test for Two Sample Means: Some General Considerations

Implications of Retaining H_0

Suppose that a different caffeine experiment yielded an obtained *t* value of 1.14 for *df* = 40. Since 1.14 is smaller than the critical value of *t* obtained from the table, you would decide to retain H_0. This does *not* imply that you have shown that the population means are equal. You do not know the probability that your decision to retain H_0 is wrong (that is, how likely you are to have made a Type II error). Unless you make use of techniques for determining the probability of a Type II error (see Chapter 14), you must limit yourself to a cautious statement such as "there is not sufficient reason to reject the hypothesis that caffeine has no effect."

"Retaining" H_0 merely means that we have not refuted it. It does *not* mean that we have shown H_0 to be true!

Implications of Rejecting H_0

When the results of the statistical analysis indicate that you should *reject H_0* (that is, the results are *statistically significant*), you may then conclude that caffeine has an effect. The probability of rejecting H_0 erroneously (committing a Type I error) *is* known; it is equal to α (or .05 in the preceding caffeine experiment). Thus, you may be 95% confident that your decision to reject H_0 is correct.

Notice, however, that there is still a 5% chance that you have made a Type I error and that the population means actually are equal. Therefore, it is usually *not* a good idea to regard any one such statistical finding as conclusive. You should wait to see if repetitions *(replications)* of the caffeine experiment also indicate that caffeine does have an effect before reaching firm conclusions.

Violations of the Underlying Assumptions

In theory, the use of the t test discussed in this chapter is justified only if two assumptions are met:

1. The variable is normally distributed within each population.
2. The variances of the two populations are equal $(\sigma_1^2 = \sigma_2^2)$.

In practice, however, you should not be deterred from using the two-tailed t test even if the assumption concerning normality is not exactly met. This test is *robust* with regard to this assumption — it gives fairly accurate results even if the assumption is not satisfied.

The assumption concerning the equality of the two population variances can also be ignored in practice *if* the two samples sizes are equal. But if the sample sizes are fairly unequal (say, the larger is more than 1.5 times greater than the smaller) *and* the population variances are markedly unequal, the t test may well yield erroneous results. So if there is reason to believe that the population variances differ substantially, it is a good idea to obtain approximately equal sample sizes. Then, any differences between the population variances will not bias the results obtained from the t test. Otherwise, there are available special tests to use when you have unequal variances with unequal sample sizes.

The *t* Test for Matched Samples

Suppose that you want to test the hypothesis that in families with two children, the first-born is more introverted than the second-born. Once again, the question of interest concerns a comparison between the means of two populations, and the null and alternative hypotheses are as follows:

$$H_0: \mu_{\text{first-born}} = \mu_{\text{second-born}}$$

$$H_1: \mu_{\text{first-born}} \neq \mu_{\text{second-born}}$$

To obtain your samples, you randomly select *matched pairs* of children, with each pair consisting of the first-born child and the second-born child from a given family. You then administer an appropriate measure of introversion. The general format of the resulting data would be as follows:

Pair	First-born (X_1)		Second-born (X_2)
1	65	matched	61
2	48	matched	42
3	63	matched	66
⋮	⋮		⋮
N	66	matched	69

The first-born child from family 1 has an introversion score of 65, and the second-born child from family 1 has an introversion score of 61. Similarly, each pair of children is matched by virtue of coming from the same family.

To illustrate a second (and perhaps more common) use of matched samples, let us suppose that you wish to test the effect of a persuasive message on people's attitudes toward gun control. Each subject's attitude is measured *before* receiving the message (X_2) and again *after* receiving the message (X_1). For pair 1, the score of 61 represents the attitude of the first subject before hearing the persuasive message, and the score of 65 represents the *same* subject's attitude after hearing the message. Each subject therefore serves as his or her own control, and each pair of scores is matched by virtue of coming from the same subject. (It would be desirable to have additional control groups in this study, but this is beyond the present point.)

The statistical analysis described previously in this chapter involved *independent* random samples. That is, there was no connection between any specific individual in sample 1 and any specific individual in sample 2. This approach is *not* suitable for matched samples. To analyze matched pair data (of either of the two types described above), procedures are used that are similar to those already developed in Chapter 10. Since each score in sample 1 has a paired counterpart in sample 2, you can subtract each X_2 from each X_1 and obtain a difference score (denoted by D). Then, you can use the techniques of the previous chapter for drawing inferences about the mean of one population—the mean of the population of difference scores.

To illustrate this procedure, let us suppose that you obtain 10 matched pairs of children, and that the results are as shown in Table 11.2. The difference scores are shown in the last column of the table; note that the *sign* of each difference score must be retained. The null and alternative hypotheses are now as follows:

$$H_0: \mu_D = 0$$

$$H_1: \mu_D \neq 0$$

H_0 states that the mean of the population of difference scores is zero, and H_1 states that the mean of the population of difference scores is not zero. The next step is to compute the mean difference score obtained from the sample, $\overline{D}$:

$$\overline{D} = \frac{\sum D}{N} = \frac{20}{10} = 2.0$$

Note that N is not the number of subjects, but the number of *pairs* of subjects. As is shown in Table 11.2, $\overline{D}$ is equal to $\overline{X}_1 - \overline{X}_2$. Thus, testing the hypothesis that $\mu_D = 0$ is equivalent to testing the hypothesis that $\mu_1 - \mu_2 = 0$, or that $\mu_1 = \mu_2$.

To test the hypothesis about the mean of one population, you compare the sample mean to the hypothesized value of the population mean and divide by the standard error of the mean, as follows:

TABLE 11.2

Introversion scores for 10 matched pairs of siblings (hypothetical data)

Pair	First-born (X_1)	Second-born (X_2)	$D = (X_1 - X_2)$
1	65	61	+4
2	48	42	+6
3	63	66	−3
4	52	52	0
5	61	47	+14
6	53	58	−5
7	63	65	−2
8	70	62	+8
9	65	64	+1
10	66	69	−3
$\sum$	606	586	+20
Mean	60.6	58.6	2.0

$$t = \frac{\overline{D} - \mu_D}{\sqrt{\dfrac{s_D^2}{N}}}$$

Here, s_D^2 is equal to the estimated population variance of the difference scores, computed by the usual formula:

$$s_D^2 = \frac{\sum (D - \overline{D})^2}{N - 1} \quad \text{or} \quad s_D^2 = \frac{\sum D^2 - \dfrac{(\sum D)^2}{N}}{N - 1}$$

Since the hypothesized value of μ_D is zero, the t formula may be simplified to:

$$t = \frac{\overline{D}}{\sqrt{s_D^2/N}}$$

These two steps may be summarized by the following (algebraically identical) computing formula:

$$t = \frac{\sum D}{\sqrt{\dfrac{N \sum D^2 - (\sum D)^2}{N - 1}}}$$

As is normally the case in tests concerning one population mean, the number of degrees of freedom on which t is based is equal to one less than the number of scores, or

$$df = (\text{number of } \textit{pairs}) - 1$$

If the computed t is smaller in absolute value than the value of t obtained from the table for the appropriate degrees of freedom, retain H_0; otherwise reject H_0 in favor of H_1.

Returning to the introversion study, the analysis of the data is as follows:

$$\alpha = .05$$

$$\sum D = 20, \quad \overline{D} = 2.0, \quad \sum D^2 = 360$$

$$s_D^2 = \frac{360 - \dfrac{(20)^2}{10}}{9} = 35.56$$

$$t = \frac{2.0}{\sqrt{35.56/10}} = 1.06$$

Or, using the computing formula,

$$t = \frac{20}{\sqrt{\dfrac{(10)(360) - (20)^2}{9}}} = 1.06$$

The critical value of t obtained from the table for $(10 - 1)$ or 9 degrees of freedom is 2.26. Since the obtained t value of 1.06 is smaller in absolute value than the critical value, you retain H_0: there is *not* sufficient reason to believe that first-born and second-born children differ in introversion.

Confidence intervals may be established in the usual way. The critical value of t is 2.26, and the standard error of the mean is equal to $\sqrt{35.56/10}$ or 1.89. The 95% confidence interval is:

$$2 - (2.26)(1.89) \leqslant \mu_D \leqslant 2 + (2.26)(1.89)$$

$$-2.27 \leqslant \mu_D \leqslant 6.27$$

As would be expected from the failure to obtain statistical significance, the value of zero falls within this interval.

Note that the analysis would be exactly the same if these data were from the gun control experiment. The calculations and test of significance would be identical, and the conclusion would be that there is *not* sufficient reason to believe that the persuasive message has any effect on people's attitudes toward gun control.

When observations can be paired or matched in a relevant way (e.g., twins, pre-post scores on the same subject), the matched t test can be a valuable experimental design. Differences between means of matched samples are smaller than they would be for independent random samples. Therefore, the standard deviation (i.e., standard error) of the distribution of differences drawn from a population of matched observations will be smaller than the standard deviation of differences drawn from independent samples. How does this affect the t test? Recall that the *smaller* the standard error, the larger the t. Therefore, any given mean population difference is likely to be more readily detected when using matched or paired samples.

Unfortunately, it is often not possible to match for various reasons: no relevant variable available, insufficient control over conditions, transfer of learning effects when using the same subject in more than one condition, etc. Therefore, while this technique is powerful, it is not always applicable.

Summary

To test null hypotheses about *differences between the means of two populations*:

1. **For independent samples,** first compute

$$s_{pooled}^2 = \frac{(N_1 - 1)\, s_1^2 + (N_2 - 1)\, s_2^2}{N_1 + N_2 - 2}$$

Then use

$$t = \frac{\overline{X}_1 - \overline{X}_2}{s_{\overline{X}_1 - \overline{X}_2}} = \frac{\overline{X}_1 - \overline{X}_2}{\sqrt{s_{pooled}^2 \left(\dfrac{1}{N_1} + \dfrac{1}{N_2} \right)}}$$

with

$$df = N_1 + N_2 - 2$$

where

$$s_{\overline{X}_1 - \overline{X}_2} = \textbf{the estimated standard error of the difference}$$

a measure of how accurate the observed difference between the sample means is likely to be as an estimate of the difference between the population means.

Reject H_0 if the computed value of t, ignoring the sign, is greater than the critical value of t obtained from the t table. Note that this t formula may well yield misleading results if σ_1^2 and σ_2^2 are markedly unequal *and* also N_1 and N_2 are markedly unequal.

2. **For matched samples,** first compute a difference score (symbolized by D) for each pair, where

$$D = X_1 - X_2$$

Next, compute the *variance* of the D scores, using the usual formula for the variance of a sample (Chapter 5). Then use

$$t = \frac{\overline{D}}{\sqrt{\dfrac{s_D^2}{N}}}$$

with

$$df = \textbf{(number of pairs)} - 1$$

where

$$\overline{D} = \text{mean of the } D \text{ scores}$$
$$s_D^2 = \text{variance of the } D \text{ scores}$$
$$N = \text{number of pairs}$$

This procedure may be summarized in the following computing formula, which yields exactly the same result:

$$t = \frac{\sum D}{\sqrt{\dfrac{N \sum D^2 - (\sum D)^2}{N - 1}}}$$

with

$$df = (\text{number of pairs}) - 1$$

Here again, reject H_0 if the computed value of t (ignoring the sign) is greater than the critical value of t obtained from the t table.

3. *Confidence intervals.*   For either independent or matched samples, confidence intervals can be established in the usual way. However, the correct standard error term must be used (the standard error of the difference with independent samples; the standard error of the mean of the differences with matched samples). Here again, confidence intervals provide considerably more information than do null-hypothesis tests.

Chapter 12
Linear Correlation and Prediction

PREVIEW

Introduction

What is meant by the correlation between two variables?

What is a positive correlation? A negative correlation?

What is a scatter plot?

How is the ability to make predictions related to correlation?

What is a linear relationship?

Describing the Linear Relationship Between Two Variables

What does the Z score difference formula for r tell us about the meaning of the correlation coefficient?

Why is it easier to determine the numerical value of r by using the computing (or raw score) formula?

If two variables are significantly correlated, does this mean that one of the variables causes the other?

How can restriction of range lead to misleading conclusions about the correlation between two variables?

Testing the Significance of the Correlation Coefficient

What are the procedures for testing hypotheses about the population correlation coefficient (usually, that it is equal to zero)?

How confident can we be if we decide to retain the null hypothesis?

How confident can we be if we decide to reject the null hypothesis?

Why do we need *both* statistical significance *and* a reasonably high absolute value of r before we can conclude that the population correlation coefficient is large enough to be useful in applied work?

If r equals zero, might there still be some relationship between the two variables?

What assumptions underly the use of r and the test for statistical significance?

Prediction and Linear Regression

What are the procedures for computing the regression line for predicting scores on Y from scores on X?

(continued next page)

Introduction

Another important objective of the behavioral scientist is to understand the relationships among different events. Such knowledge enables us to make accurate predictions about future events, and these predictions make it possible for us to achieve some degree of control over our formidable environment.

For example, you learn early in your scholastic career that the amount of time spent in studying is related to grades. True, there are exceptions: some students may study for many hours and obtain poor grades, while others may achieve high grades despite short study periods. Nevertheless, the general trend holds. In the majority of cases, you can accurately predict that little or no studying will lead to poor grades, while more studying will result in better grades. So if you are dissatisfied with grades that you feel are too low, one likely way to improve this aspect of your environment is by studying more. In statistical terminology, the two variables of hours studying and grades are said to be co-related or *correlated.*

Many pairs of variables are correlated, while many are unrelated or *uncorrelated.* For example, sociologists have found that the incomes of families are positively related to the IQs of the children in the families; the more income, the higher the children's IQs. These variables are said to be *positively correlated.* This relationship is depicted graphically in Figure 12.1, which is called a *scatter plot* (or scatter diagram) because the points scatter across the range of scores. Note that each point on the graph represents two values for one family, income (X variable) and IQ of the child (Y variable). Also, in the case of a positive correlation, the straight line summarizing the points slopes *up* from left to right.

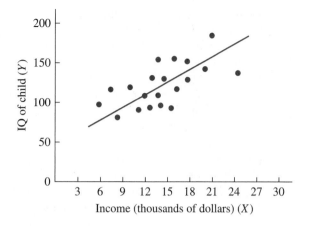

FIGURE 12.1

Relationship between income of a family and IQ of a child in that family

The golf enthusiast will readily agree that the number of years of play is negatively related to his golf score; the more years of practice, the fewer shots needed to complete a round of 18 holes. These two variables are said to be *negatively correlated*. (See Figure 12.2.) Notice that in the case of a

FIGURE 12.2

Relationship between years of play and average score for ten golfers

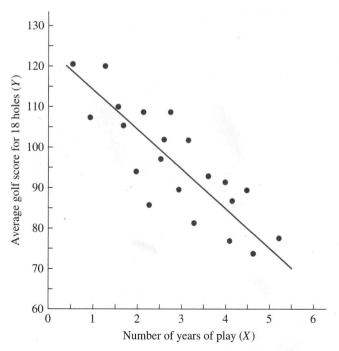

negative correlation, the straight line summarizing the points slopes *down* from left to right.

A different picture results when two variables are uncorrelated. For example, length of big toes among male adults is uncorrelated with IQ scores, and the corresponding scatter plot is shown in Figure 12.3.

When you can demonstrate that two variables are correlated, you can use the score of an individual on one variable to *predict* or *estimate* his score on the other variable. In Figure 12.2, a fairly accurate prediction of an individual's golf score can be made from the number of years he has played golf. The more closely the two variables are related, the better the prediction is likely to be; while if two variables are uncorrelated (as in the case of toe length and IQ), you cannot accurately predict an individual's score on one of them from his score on the other. Thus the concepts of correlation and prediction are closely related.

This chapter deals with three main topics: (1) the *description* of the relationship between two variables for a set of observations; (2) making *inferences* about the strength of the relationship between two variables in a population, given the data of a sample; and (3) the *prediction* or estimation of values on one variable from observations on another variable with which it is paired. We will deal with one very common kind of relationship between two variables, namely a *linear* or straight-line relationship. That is, if values of one variable are plotted against values of the other variable on a graph, the trend of the plotted points is well represented by a straight line.

FIGURE 12.3

Relationship between the length of big toe and IQ scores for adult men

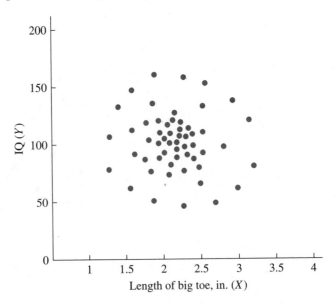

Notice that the data plotted in Figures 12.1 and 12.2 tend to fall near or on the straight line drawn through the scatter plot; this indicates that the two variables in question are highly linearly correlated. The points in Figure 12.3, on the other hand, are scattered randomly throughout the graph and cannot be represented well by a straight line. Therefore, these two variables are linearly uncorrelated.

Many pairs of variables that are of importance to behavioral scientists tend to be linearly related. Although there are many other ways of describing relationships, the linear model is generally the one most frequently used in such fields as psychology, education, and sociology.

Describing the Linear Relationship Between Two Variables

Suppose you want to measure the degree of linear relationship between a scholastic aptitude test (SAT) and college grade point average (GPA). It would be reasonable to expect these two variables to be positively correlated. This means that students with high SAT scores, on the average, obtain relatively high GPAs and students with low SAT scores, on the average, obtain low GPAs. The phrase "on the average" alerts us to the fact that there will be exceptions: some students with low SAT scores will do well in college and obtain high GPAs, while some with high SAT scores will do poorly and receive low GPAs. That is, the relationship between SAT scores and GPA is not perfect. Thus the question arises: Just how strong *is* the relationship? How can it be summarized in a single number?

The *Z* Score Difference Formula for *r*

We have seen in Chapter 6 that in order to compare scores on different variables (such as a mathematics test and a psychology test), it is useful to transform the raw scores into standard scores. These transformed scores allow you to compare paired values directly. Similarly, in order to obtain a coefficient of relationship which describes the similarity between paired measures in a single number, the raw score must be transformed into Z units:

$$Z_X = \frac{X - \overline{X}}{\sigma_X} \qquad\qquad Z_Y = \frac{Y - \overline{Y}}{\sigma_Y}$$

In Table 12.1, a distribution of SAT scores (X) and GPAs (Y) is presented along with the corresponding descriptive statistics for 25 students at a college in the eastern United States. The SAT scores were obtained when the students were high-school seniors, and the GPAs are those received by the students after one year of college. In addition to the raw scores, the Z equivalents are also shown in the table. Notice that students with high SAT scores do tend to have high GPAs, and consequently large Z_X and Z_Y values.

TABLE 12.1

Raw scores and Z scores on SAT (X) and GPA (Y) for 25 students in an eastern U.S. college

Student	X	Y	Z_X	Z_Y
1	650	3.8	1.29	1.67
2	625	3.6	1.08	1.31
3	480	2.8	−.16	−.09
4	440	2.6	−.50	−.44
5	600	3.7	.86	1.49
6	220	1.2	−2.37	−2.89
7	640	2.2	1.21	−1.14
8	725	3.0	1.93	.26
9	520	3.1	.18	.44
10	480	3.0	−.16	.26
11	370	2.8	−1.09	−.09
12	320	2.7	−1.52	−.26
13	425	2.6	−.62	−.44
14	475	2.6	−.20	−.44
15	490	3.1	−.07	.44
16	620	3.8	1.04	1.67
17	340	2.4	−1.35	−.79
18	420	2.9	−.67	.09
19	480	2.8	−.16	−.09
20	530	3.2	.27	.61
21	680	3.2	1.55	.61
22	420	2.4	−.67	−.79
23	490	2.8	−.07	−.09
24	500	1.9	.01	−1.67
25	520	3.0	.18	.26
$\overline{X} = 498.4$		$\overline{Y} = 2.85$	$\sigma_X = 117.33$	$\sigma_Y = .57$

Conversely, students with low SAT scores tend to have low GPAs. So the paired Z values are similar for most students, and the two variables are therefore highly positively correlated. The *raw scores,* however, need not be numerically similar for any pair, since they are measured in different units for each variable.

If the association between the two selected variables were perfect and in a positive direction, each person would have exactly the same Z_X and Z_Y paired values. If the relationship were perfect but in a negative direction, each Z_X value would be paired with an identical Z_Y value but they would be *opposite in sign.* (See Figure 12.4.) These perfect relationships are offered for illustration and almost never occur in practice.

Since the size of the difference between paired Z values is related to the amount of the relationship between the two variables, some kind of *average* of these differences should yield information about the closeness or *strength* of the association between the variables. Because the mean of the differences $Z_X - Z_Y$ is necessarily zero, the coefficient of correlation is actually

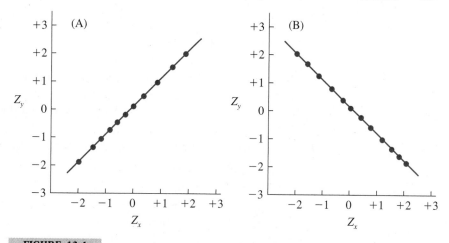

FIGURE 12.4

Perfect linear relationship between two variables: (A) Perfect positive linear relationship; (B) perfect negative linear relationship

obtained by *squaring* differences between paired Z values. The size of the average of these squared differences,

$$\frac{\sum (Z_X - Z_Y)^2}{N}$$

is an index of the strength of the relationship. For example, a small value indicates a high positive correlation (little difference between the paired Z values). If this average is a large number, it indicates a high negative relationship (most paired Z values "opposite" to one another). A "medium size" average indicates little or no relationship. This average, however, is not convenient to interpret, since (as can be proved) it ranges from zero in the case of a perfect positive correlation to 4.0 in the case of a perfect negative relationship. A much more readily interpretable coefficient is obtained by subtracting half of this average from one:

$$r_{XY} = 1 - \frac{1}{2} \frac{\sum (Z_X - Z_Y)^2}{N}$$

where

r_{XY} = **symbol for the correlation coefficient between** X **and** Y
N = **number of *pairs***

(It should be noted that this is *not* a conventional computing formula.)

This correlation coefficient has the following desirable characteristics:

1. A value of zero indicates no linear relationship between the two variables (that is, they are linearly uncorrelated).

2. The *size* of the numerical value of the coefficient indicates the *strength* of the relationship. Large absolute values mean that the two variables are closely related, and small absolute values mean that they are only weakly related.

3. The *sign* of the coefficient indicates the *direction* of the relationship.

4. The largest possible positive value is $+1.00$, and the largest possible negative value is -1.00.

Thus, if the correlation between two variables is $+.20$, you can tell at a glance that the relationship between them is positive and weak (since it is far from the maximum of $+1$). A correlation of $-.80$ would indicate a strong negative relationship.

The correlation coefficient may be symbolized more simply by r when there is no possible confusion as to which two variables are involved. This coefficient is often referred to as the Pearson r in honor of Karl Pearson, who did the early work on this measure starting with an idea of Francis Galton.

To clarify the meaning of the above formula, let us suppose that we have a small set of four paired X and Y scores, obtained from two quizzes in a small college seminar. (Four pairs would be far too few to permit a useful conclusion in a real experiment, but it will be easier to illustrate the concept if very few numbers are involved.) Let us first suppose that the results are as follows:

Student	X	Y	Z_X	Z_Y	$Z_X - Z_Y$
1	8	29	$+1.266$	$+1.266$	0
2	7	26	$+.633$	$+.633$	0
3	5	20	$-.633$	$-.633$	0
4	4	17	-1.266	-1.266	0

Note that the paired Z values are identical in each case, indicating a perfect positive relationship. There is no difference between any of the corresponding Z values, so the correlation coefficient is equal to

$$r = 1 - \frac{1}{2}\frac{(0)^2 + (0)^2 + (0)^2 + (0)^2}{4}$$

$$\doteq 1 - \frac{1}{2}(0)$$

$$= +1.00$$

Thus the smallest possible mean of $(Z_X - Z_Y)^2$ is 0. So the largest possible positive value of r is $+1$, the value that occurs when the relationship between the two variables is perfect and positive.

Now let us suppose that the results are instead as follows:

Student	X	Y	Z_X	Z_Y	$Z_X - Z_Y$
1	8	17	$+1.266$	-1.266	$+2.532$
2	7	20	$+.633$	$-.633$	$+1.266$
3	5	26	$-.633$	$+.633$	-1.266
4	4	29	-1.266	$+1.266$	-2.532

In this case, the relationship between the two variables is perfect and negative. This is indicated by the fact that the Z values for each person are equal but opposite in sign. In other words, high scores on X are paired with low scores on Y, and low scores on X are paired with high scores on Y. The correlation coefficient is equal to

$$r = 1 - \frac{1}{2} \frac{(+2.532)^2 + (+1.266)^2 + (-1.266)^2 + (-2.532)^2}{4}$$

$$= 1 - \frac{1}{2}\left(\frac{16}{4}\right)$$

$$= 1 - \frac{1}{2}(4)$$

$$= 1 - 2$$

$$= -1.00$$

Thus the largest possible value of the mean of $(Z_X - X_Y)^2$ is 4, as stated above. So the largest possible negative value of r is -1, the value that occurs when the relationship between the two variables is perfect and negative.

Finally, let us suppose that the results are instead as follows:

Student	X	Y	Z_X	Z_Y	$Z_X - Z_Y$
1	8	20	$+1.266$	$-.633$	$+1.899$
2	7	29	$+.633$	$+1.266$	$-.633$
3	5	17	$-.633$	-1.266	$+.633$
4	4	26	-1.266	$+.633$	-1.899

Here, there is no linear relationship at all between X and Y, as is shown by the paired Z values, and the correlation coefficient is equal to

$$r = 1 - \frac{1}{2} \cdot \frac{(+1.899)^2 + (-.633)^2 + (+.633)^2 + (-1.899)^2}{4}$$

$$= 1 - \frac{1}{2} \left(\frac{8}{4} \right)$$

$$= 1 - \frac{1}{2} (2)$$

$$= 1 - 1$$

$$= 0$$

The correlation coefficient can take on any value between -1.00 and $+1.00$. Equal numerical values of r describe equally strong relationships between variables. For example, coefficients of $+.50$ and $-.50$ describe relationships which are equally strong but are in opposite directions.

Computing Formulas for r

The preceding procedure for obtaining r, although best for understanding the meaning of r, is much too tedious computationally even if calculators are available. It can be shown,* however, that identical results are obtained by calculating the mean of the *product* of the paired Z values:

$$r_{XY} = \frac{\sum Z_X X_Y}{N}$$

We can avoid the necessity for converting to Z values by rewriting the preceding formula as

$$r_{XY} = \frac{\sum (X - \overline{X})(Y - \overline{Y})}{N \sigma_X \sigma_Y}$$

When further spelled out, this gives us the *raw score formula for the Pearson correlation coefficient:*

$$r_{XY} = \frac{N \sum XY - \sum X \sum Y}{\sqrt{[N \sum X^2 - (\sum X)^2][N \sum Y^2 - (\sum Y)^2]}}$$

While the raw score formula looks formidable, it is the easiest to use in practice. Note that the formula calls for both $\sum XY$ and $\sum X$ times $\sum Y$, which are *not* the same (see Chapter 1), and that N in all of the above formulas stands for the number of *pairs* of observations (the number of cases).

---

* Proofs of the equivalence of all of the various formulas for r are given in the Appendix at the end of this chapter.

The calculation of the Pearson r for the SAT and GPA data in Table 12.1, using both the raw score and Z product formula, is shown in Table 12.2. The obtained value of $+.65$ indicates that for this group of 25 students, there is a

TABLE 12.2

Calculation of Pearson correlation coefficient between SAT (X) and GPA (Y) by the raw score and Z product methods

Subject	\multicolumn{5}{c}{Raw Score Method}	\multicolumn{3}{c}{Z Product Method}
	X	Y	XY	X^2	Y^2	Z_X	Z_Y	$Z_X Z_Y$
1	650	3.8	2470	422500	14.44	1.29	1.67	2.1543
2	625	3.6	2250	390625	12.96	1.08	1.31	1.4148
3	480	2.8	1344	230400	7.84	$-.16$	$-.09$	.0144
4	440	2.6	1144	193600	6.76	$-.50$	$-.44$	.2200
5	600	3.7	2220	360000	13.69	.86	1.49	1.2814
6	220	1.2	264	48400	1.44	-2.37	-2.89	6.8493
7	640	2.2	1408	409600	4.84	1.21	-1.14	-1.3794
8	725	3.0	2175	525625	9.00	1.93	.26	.5018
9	520	3.1	1612	270400	9.61	.18	.44	.0792
10	480	3.0	1440	230400	9.00	$-.16$	.26	$-.0416$
11	370	2.8	1036	136900	7.84	-1.09	$-.09$	.0981
12	320	2.7	864	102400	7.29	-1.52	$-.26$	.3952
13	425	2.6	1105	180625	6.76	$-.62$	$-.44$	.2728
14	475	2.6	1235	225625	6.76	$-.20$	$-.44$	.0880
15	490	3.1	1519	240100	9.61	$-.07$	.44	$-.0308$
16	620	3.8	2356	384400	14.44	1.04	1.67	1.7368
17	340	2.4	816	115600	5.76	-1.35	$-.79$	1.0665
18	420	2.9	1218	176400	8.41	$-.67$	.09	$-.0603$
19	480	2.8	1344	230400	7.84	$-.16$	$-.09$	.0144
20	530	3.2	1696	280900	10.24	.27	.61	.1647
21	680	3.2	2176	462400	10.24	1.55	.61	.9455
22	420	2.4	1008	176400	5.76	$-.67$	$-.79$	.5293
23	490	2.8	1372	240100	7.84	$-.07$	$-.09$	.0063
24	500	1.9	950	250000	3.61	.01	-1.67	$-.0167$
25	520	3.0	1560	270400	9.00	.18	$-.26$	.0468

$\sum X = 12{,}460 \qquad (\sum X)^2 = (12{,}460)^2$

$\sum Y = 71.2 \qquad\qquad\qquad = 155{,}251{,}600$

$\sum XY = 36{,}582$

$\sum X^2 = 6{,}554{,}200 \qquad (\sum Y)^2 = (71.2)^2$

$\sum Y^2 = 210.98 \qquad\qquad\quad = 5069.44$

$N = 25$

$\sum Z_X Z_Y = 16.35$

$N = 25$

$$r = \frac{\sum Z_X Z_Y}{N} = \frac{16.35}{25} = +.65$$

$$r = \frac{N \sum XY - \sum X \sum Y}{\sqrt{[N \sum X^2 - (\sum X)^2][N \sum Y^2 - (\sum Y)^2]}}$$

$$r = \frac{(25)(36{,}582) - (12{,}460)(71.2)}{\sqrt{[25(6{,}554{,}200) - 155{,}251{,}600][25(210.98) - 5069.44]}}$$

$$r = +.65$$

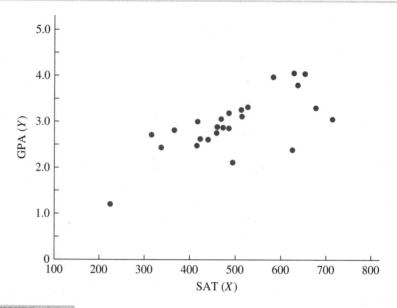

FIGURE 12.5
Scatter plot for data in Table 12.1 $(r = +.65)$

high positive correlation between these two variables. (We call this value "high" because values for r between different variables which are as large as this are rare in the fields where r is used.) The corresponding scatter plot is shown in Figure 12.5.

Correlation and Causation

Some cautions must be noted at this point. First, you cannot determine the *cause* of the relationship from the correlation coefficient. Two variables may be highly correlated for one or more of three reasons: (1) X causes Y, (2) Y causes X, or (3) both X and Y are caused by some third variable.

A well-known story illustrates the danger of inferring causation from a correlation coefficient. A researcher once obtained a high positive correlation between the number of storks and the number of births in European cities (that is, the more storks in a city, the more births). Instead of issuing a dramatic announcement supporting the mythical powers of storks, further investigation was carried out. It was found that storks nest in chimneys, which in turn led to the conclusion that a third variable was responsible for the relationship between storks and births — size of city. Large cities had more people, and hence more births. They also had more houses, and hence more chimneys, and hence more storks. Smaller cities had fewer people, births, houses, chimneys, and storks. Thus attributing causality is a logical or scientific problem, not a statistical one.

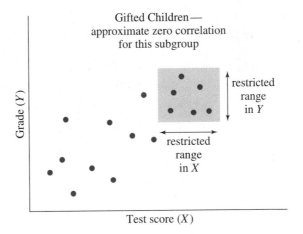

Test score (X)

FIGURE 12.6
Effect of restriction of range on the correlation coefficient

Correlation and Restriction of Range

A second important point has to do with the effect of the variability of the scores on the correlation coefficient. Suppose you calculate the correlation between achievement test scores and elementary school grades for public school children in New York City. You will probably find a strong linear trend when the data are plotted. But what if you were to calculate the correlation between the same two variables for a group of children in a class for the highly gifted? Here, the range of scores on both variables is considerably narrowed, as is shown in Figure 12.6. Because of this, the correlation between the two variables is markedly reduced.

It is much harder to make fine discriminations among cases that are nearly equal on the variables in question than it is to make discriminations among cases that differ widely. More specifically, it is difficult to predict whether a gifted child will be an A or A− student. It is much easier to distinguish among a broad range of A to F students. Thus, *when the variability of scores is restricted* on either or both of two variables, the correlation between them decreases in absolute value. Conversely, when the variability of scores is increased on either of two variables (for example, by dropping cases in the middle of the range), the correlation can be expected to increase in absolute value.

Testing the Significance of the Correlation Coefficient

The correlation coefficient of +.65 for the group of 25 students whose scores are shown in Table 12.1 conveniently describes the linear relationship between SAT scores and GPA *for this group*. It would be very useful to know, however, whether these two variables are correlated in the population of all students. To answer this question, we must draw an inference about likely values of the population correlation coefficient (symbolized by rho, ρ). Is it likely that there is actually no correlation in the population, and that the correlation in the sample of 25 students was due to sampling error (the cases that happened to wind up in the sample)? Or, is the value of .65 large enough for us to conclude that there is some nonzero positive correlation between SAT scores and GPA in the population?

The strategy for testing hypotheses about likely values of ρ is similar to that used in previous chapters. The null hypothesis most often tested is that ρ is equal to zero:*

$$H_0: \rho = 0$$

$$H_1: \rho \neq 0$$

A criterion of significance, such as the .05 or .01 criterion, is selected. The correct statistical model to use in this situation is the t distributions, so the appropriate t ratio may now be computed† with degrees of freedom equal to $N - 2$. Then, H_0 is retained if the computed t is less than the critical value of t from the t table. Otherwise H_0 is rejected in favor of H_1, and r is said to be *significantly different from zero* (or, simply, *statistically significant*).

However, you do not necessarily need to compute the t ratio. This has (in a sense) already been done for you by statisticians who have constructed tables of significant values of r. Therefore, the procedure for testing a correlation coefficient for statistical significance can be quite simple: you compare your computed value of r to the value of r shown in Table D in the Appendix for $N - 2$ degrees of freedom, where N is equal to the number of pairs. If the absolute value of your computed r is smaller than the tabled value, retain H_0; otherwise, reject H_0.

As an illustration, suppose that you wish to test the significance of the correlation between SAT scores and GPAs, using $\alpha = .05$. Referring to Table D, you find that for $25 - 2$ or 23 degrees of freedom, the smallest absolute value of r that is statistically significant is .396. Since the obtained r of .65 exceeds this value, you reject H_0 and conclude that there *is* a positive correlation in the population from which the sample of 25 students was selected. Note that a correlation of $-.65$ would indicate a statistically significant negative relationship; the sign of r is ignored when comparing the computed r to the tabled value of r. Thus the test described is two-tailed—it guards

---

* Other null hypotheses are possible, but a different statistical procedure is required in order to test them. See, for example, W. L. Hays, *Statistics*, 3rd ed. (New York: Holt, 1981), 464–467.

† The formula for the t ratio is

$$t = \frac{r\sqrt{N - 2}}{\sqrt{1 - r^2}}$$

where N = number of *pairs* of scores.

against chance departures of r from zero in either the positive or negative direction.

As we observed in Chapters 10 and 11, confidence intervals provide valuable additional information about the probable values of a population parameter. However, the procedures for calculating confidence intervals for correlation coefficients are beyond the scope of this book.*

Implications of Retaining H_0

If you fail to reject H_0, you have *not* established that the two variables are linearly uncorrelated in the population. Unless you know the probability of a Type II error (see Chapter 14), you cannot tell to what extent your decision to retain H_0 is likely to be wrong. Therefore, you should limit yourself to a conservative statement such as "there is not sufficient reason to believe that ρ is different from zero."

Also, the Pearson r detects only linear relationships. Thus the possibility remains that the two variables are related, but not in a linear fashion. It is possible for two variables to be even perfectly related, yet to have r equal to zero. In Figure 12.7, for example, you can predict Y without error given X, but *not* linearly.

Implications of Rejecting H_0

If a Pearson r is statistically significant, the significance denotes *some* degree of linear relationship between the two variables in the population. It does *not* denote a "significantly high" or "significantly strong" relationship; it just means that the relationship in the population is unlikely to be zero.

Notice that the larger the sample size, the smaller the absolute value of the correlation coefficient that is needed for statistical significance. For example, for $N = 12$ ($df = 10$), a correlation of .576 or larger is needed for significance using the .05 criterion. For $N = 102$ ($df = 100$), a correlation of

FIGURE 12.7

Example of perfect relationship between two variables where $r = 0$

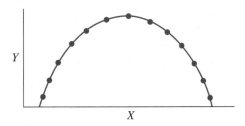

---

*A detailed presentation may be found in J. Cohen & P. Cohen, *Applied multiple regression/correlation analysis for the behavioral sciences*, 2nd ed. (Hillsdale, N.J.: Lawrence Erlbaum Associates, 1983).

only .195 or larger is needed. This implies that the importance of obtaining statistical significance can easily be exaggerated. Suppose that for a very large sample (say, $N = 1000$), a statistically significant Pearson r of .08 is obtained. This statistical test indicates that ρ is greater than zero. But the obtained r of .08 is so close to zero that the relationship in the population, although not zero and not necessarily equal to .08, is likely to be *very* weak—too weak, in fact, to allow you to make any accurate statements about one variable based on knowledge of the other. Consequently, such a finding would add little or nothing to our immediate knowledge, even though it is statistically significant. (It might conceivably be useful in appraising a theory.) Therefore, you need to have *both* statistical significance *and* a reasonably high absolute value of r before you can conclude that ρ is large enough to be useful in applied work.

Correlation coefficients *cannot* be interpreted as percents. For example, you *cannot* conclude that a correlation of .60 is 60% of a perfect relationship or that it is twice as much as a correlation of .30. As we will see later in this chapter, however, the *squared* correlation coefficient (r^2) does permit an interpretation, in percentage terms, of the strength of the relationship between the two variables.

As usual, given that H_0 is true, the probability of erroneously rejecting it (a Type I error) is equal to the criterion of significance that is selected. Using $\alpha = .05$, when H_0 is true, you will reject it 5% of the time.

Assumptions Underlying the Use of *r*

The most important assumption underlying the use of the Pearson r is that X and Y are linearly related. If X and Y have a curvilinear relationship (as in Figure 12.7), the Pearson r will not detect it. Even this is not, strictly speaking, an assumption. If r is considered a measure of the degree of *linear* relationship, it remains such a measure whether or not the best-fitting function is linear. Ordinarily, however, one would not be interested in the best linear fit when it is known that the relationship is not linear.

Otherwise, no assumptions are made at all in using r to *describe* the degree of linear relationship between two variables for a given set of data. When testing a correlation coefficient for statistical significance, it is, strictly speaking, assumed that the underlying distribution is the so-called bivariate normal—that is, scores on Y are normally distributed for each value of X, and scores on X are normally distributed for each value of Y. However, when the degrees of freedom are greater than 25 or 30, failure to meet this assumption has little consequence to the validity of the test.

Prediction and Linear Regression

Behavioral scientists are indebted to Sir Francis Galton for making explicit some elementary concepts of relationships and prediction. Galton wrote his now classic paper in 1885, *Regression toward mediocrity in hereditary stature.*

There he presented the theory that the physical characteristics of offspring tend to be related to, but are on the average less extreme than, those of their parents. That is, tall parents on the average produce children *less tall* than themselves, and short parents on the average produce children *less short* than themselves. In other words, physical characteristics of offspring tend to "regress" toward the average of the population. If you were to predict the height of a child from a knowledge of the height of the parents, you should predict a less extreme height — one closer to the average of all children.

Plotting data on the stature of many pairs of parents and offspring, Galton calculated the median height of offspring for each "height" category of parents. For example, he plotted the median height for all offspring whose fathers were 5 ft 7 in., the median height for all offspring whose fathers were 5 ft 8 in., and so forth. By connecting the points representing the medians, Galton found not only that there was a positive relationship between parental height and height of the offspring, but also that this relationship was fairly linear. The line connecting each of the medians (and after Pearson, the means) came to be known as the *regression line*. This term has been adopted by statisticians to indicate the straight line used in predicting or estimating values of one variable from a knowledge of values of another variable with which it is paired. The statistical procedure for making such predictions is called *linear regression.*

Rationale and Computational Procedures

Let us assume that we wish to predict scores on Y from scores on X. Y, the variable being predicted or estimated, is called the *dependent* variable or *criterion*. X, which provides the information on which the predictions are based, is referred to as the *independent* variable or *predictor*. In this book, we will discuss the simplest method of prediction — an equation that produces a *straight line*. One way of writing such an equation is

$$Y' = b_{YX}X + a_{YX}$$

where

$Y' = predicted$ score on Y
$b_{YX} = slope$ of the line, also called the *regression coefficient*
for predicting Y
$a_{YX} = Y\text{-}intercept$, or the value of Y' when $X = 0$

The value of b_{YX} summarizes the average rate of change in Y score per unit increase in X score, while the value of a_{YX} indicates the Y value at which the line crosses the Y-axis. For any given set of data, b_{YX} and a_{YX} are constant values in the equation. A straight line has only one slope, and there is only one value of Y' for which X equals zero. We wish to keep our errors in prediction to a minimum, and the actual values of b_{YX} and a_{YX} are chosen with this objective in mind.

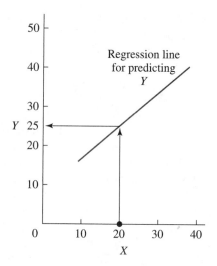

Use of regression line to obtain predicted scores on Y

The way in which a regression line is used to obtain a predicted score is illustrated in Figure 12.8. Suppose a person has a score of 20 on X, and you want to predict what her score on Y will be. Given the regression line for predicting Y shown in Figure 12.8, you enter the X axis at 20 and proceed up to the regression line. You then read out the predicted Y score; Y' for this subject is equal to 25. In practice, predicted scores are computed according to the regression equation, since graphic estimates are likely to be rather inaccurate.

The regression (or prediction) line is the straight line that best represents the trend of the dots in a scatter plot. In any real situation, the dots will *not* all fall exactly in a straight line. Therefore, you will make *errors* when you use a regression line to make predictions. The error in predicting a particular Y value is defined as the difference (keeping the sign) between the *actual Y* value and the *predicted Y* value, Y'. That is,

$$\text{error in predicting } Y = Y - Y'$$

A graphic representation of errors in prediction for one set of data is shown in Figure 12.9. For example, the predicted Y value for the individual with an X score of 68 is equal to approximately 159.8. His actual Y value, however, is equal to 148. The difference of approximately -11.8 between the actual and predicted values represents an error in prediction for this individual, and the negative sign indicates that the actual value is smaller than the prediction.

In Figure 12.9, vertical lines have been drawn between each observed Y score (actual weight) and the predicted weight score (Y'). This has been done for each value of X (actual height). These vertical distances are the er-

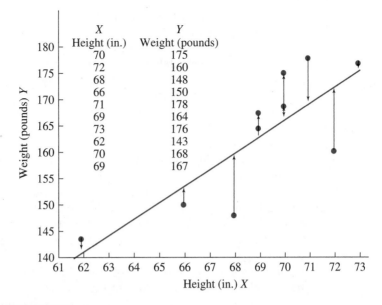

X	Y
Height (in.)	Weight (pounds)
70	175
72	160
68	148
66	150
71	178
69	164
73	176
62	143
70	168
69	167

FIGURE 12.9

Regression line of Y on X showing extent of error (difference between actual weight score and predicted weight score)

rors in prediction. All *predicted* scores lie on the straight line. Since there are two height scores of 70 inches with two different weight scores, and two height scores of 69 inches with two different weight scores, the smaller error has been superimposed upon the larger in each of these cases.

The linear regression prediction line minimizes the sum of *squared* errors in prediction, symbolized as

$$\sum (Y - Y')^2$$

That is, values are chosen for b_{YX} and a_{YX} which define a special regression line for predicting Y: the line which makes $\sum (Y - Y')^2$ smaller than you would get using any other prediction line for that set of data. This line is therefore called the *least-squares* regression line of Y on X.

To see why the sum of *squared* errors is minimized, rather than the sum of unsquared errors, let us suppose that you wish to predict students' college GPAs from scores on the SATs. After making your predictions, you wait until the students finish college and see what their actual GPAs turn out to be. The results for two students are:

Student	Actual GPA (Y)	Predicted GPA (Y')	Error $(Y - Y')$
1	3.8	3.4	$+0.4$
2	2.1	2.5	-0.4

The actual GPA of the first student is about half a grade point above the prediction, while the GPA of the second student is a similar amount below the prediction for that student. Thus some error in prediction has been made. (This is unavoidable unless $r = \pm 1$, which never happens with real data.) Yet if you were to compute the average of the error column, it would equal zero! The same misleading result would be obtained had the predicted Y' values been 2.4 for student 1 and 3.5 for student 2, yielding substantial errors of $+1.4$ and -1.4.

As this example shows, it is not sufficient to have a prediction line whose errors balance out (have a mean of zero); *all* lines which go through the point $\overline{X}, \overline{Y}$ have this property. In order to obtain a meaningful measure of the total amount of prediction error, we must find the line about which the variation of the Y values, and hence the *squared* errors, is as small as possible. (As we will see later in this chapter, a good measure of the magnitude of prediction error can be computed by averaging these squared errors and taking the square root.)

In the general formula for a straight line, therefore, the values of b_{YX} and a_{YX} are chosen so as to minimize the value of $\sum (Y - Y')^2$. Equations that produce these values are found (by means of the differential calculus) to be

$$b_{YX} = r_{XY} \frac{\sigma_Y}{\sigma_X} = \frac{N \sum XY - \sum X \sum Y}{N \sum X^2 - (\sum X)^2}$$

$$a_{YX} = \overline{Y} - b_{YX} \overline{X}$$

The first formula for b_{YX} is useful when values of r, σ_X, and σ_Y are available. The second formula for b_{YX} is preferable when working from raw scores.

To illustrate, let us return once again to our example of SAT and GPA. We have already seen that $\overline{X} = 498.4$, $\overline{Y} = 2.85$, $\sigma_X = 117.33$, $\sigma_Y = .57$, and $r_{XY} = +.65$, so

$$b_{YX} = (+.65) \frac{.57}{117.33}$$

$$= .0032$$

Since b_{YX} is the slope of the line, this result indicates that each unit of increase in X is associated with .0032 units of increase in Y. While it is not obvious from the equation, the value of b_{YX} does not change even if σ_X changes, while r_{XY} does. Thus, the rate of change in Y units per X remains the same whether the range of X is wide or narrow.

$$a_{YX} = 2.85 - (0.0032)(498.4)$$

$$= 1.26$$

This is the value of Y' when X is equal to zero. Such a value need not be a *logical* possibility for a given set of data. This is the case here, since SAT scores of zero are outside the observed range.

Combining the above two results gives the linear regression equation for predicting Y (GPA):

$$Y' = b_{YX} X + a_{YX}$$

$$= .0032X + 1.26$$

This equation can now be used to obtain predicted GPAs (Y'), given a SAT score (X). For example, the predicted GPA for a student with a SAT score of 400 is:

$$Y' = (.0032)(400) + 1.26 = 2.5$$

This is the *best* linear prediction you can make, the one that on the average would be least in error.

Note that a sample with scores on *both* the X and Y variables is needed in order to compute a linear regression prediction equation. Thus, to predict GPA from SAT, you must first obtain a sample of college students who have scores on both the SAT and GPA and compute such essential values as r_{XY}. Once you have determined the regression line, you can then use it to *make predictions for new cases for whom you have data on only the predictor*—that is, to make predictions for graduating high school seniors based on their SAT scores. However, you *must* be careful to ensure that the original sample of college students upon whom the regression equation is calculated is representative of the future groups for whom the predictions will be made.

Properties of Linear Regression

The linear regression procedure has numerous important properties. We have already seen that there is precisely one predictor and one criterion, a straight line is used to make predictions, and the line is such that the sum of squared errors in prediction is minimized. Some additional principles of importance include the following:

1. *If there is no good information on which to base a prediction, the same estimate—the mean of the criterion—is made for everyone.* Suppose that a group of graduating high-school seniors asks you to predict what each one's college GPA will be, but the only information that they give you is the length of each student's right big toe. Since big toe length is useless for purposes of forecasting someone's GPA (that is, $r_{XY} = 0$), you should refuse to make any predictions in this situation.

   When there is no good information on which to base a prediction, linear regression also refuses to forecast any differences on the criterion. Instead, it "predicts" that each person will be average. That is, when the correlation between the predictor and the criterion is zero, the linear regression formula becomes

$$b_{YX} = .00 \frac{\sigma_Y}{\sigma_X} = 0$$

$$a_{YX} = \overline{Y} - (.00)\overline{X} = \overline{Y}$$

$$Y' = b_{YX} X + a_{YX}$$

$$= (0)X + \overline{Y}$$

$$= \overline{Y}$$

Thus scores on X are ignored, since they are irrelevant for purposes of predicting Y, and the mean of the criterion is predicted for everyone. Of course, no behavioral scientist uses linear regression unless the predictor and criterion are nontrivially correlated, but it is desirable to understand this "worst of all possible worlds" in order to follow the logic of linear regression.

2. *When all scores are expressed as Z scores, the predicted Z score on Y is equal to r multiplied by the Z score on X.* It can be shown algebraically that*

$$Z'_Y = r_{XY}Z_X$$

where

$$Z'_Y = \text{predicted } Z \text{ value on } Y$$

$$Z_X = \text{actual } Z \text{ value on } X$$

This equation may prove helpful in understanding linear regression. It shows that the predicted score (Z'_Y) will be less extreme (that is, closer to its mean) than the score from which the prediction is made (Z_X), because Z_X is multiplied by a fraction (r_{XY}). This illustrates the statement, made previously, about regression toward the mean of Y. The equation also shows once again that if r_{XY} is equal to zero, all predicted Z'_Y values will equal the mean of Y (they will equal zero, which is the mean of Z scores). This incidentally shows that the mean is a least squares measure — that is, the sum of squared deviations of the values in the sample from it is a minimum. Finally, the equation indicates that the regression line passes through the point $(\overline{X}, \overline{Y})$. If $Z_X = 0$ (the mean of the X scores expressed as Z scores), then $Z'_Y = 0$ (the mean of the Y scores expressed as Z scores).

Although this equation looks simple, it is not in general convenient for calculating predicted scores. To use it for this purpose, you would first

---

* When the scores are expressed in deviation units, the standard deviation of both Z_X and Z_Y is equal to 1. Therefore, $b_{YX} = r(1/1) = r$. Since the mean of both Z_X and Z_Y is equal to zero, $a_{YX} = 0$. Thus we have $Y' = rX + 0$, or $Z'_Y = r_{XY}Z_X$.

have to transform X to a Z value, then compute the value of Z'_Y, and then transform Z'_Y to a raw score equivalent Y'. Most of the time, it will be easier to use the raw score regression equation.

3. *The closer the correlation between X and Y is to zero, the greater is the amount of prediction error.* It should be emphasized that while linear regression *minimizes* the sum of squared errors in prediction, this sum may still be prohibitively large. If the correlation between the predictor and the criterion is numerically small, such as $+.07$, even linear regression will make a substantial amount of prediction error because the relationship between the two variables is so weak. This implies that a correlation near zero is likely to be useless for practical prediction purposes, even if it is statistically significant (as could happen with a very large sample). The *sign* of the correlation coefficient, however, is *not* related to prediction error. A correlation of (say) $+.55$ and one of $-.55$ are in theory equally good for prediction purposes, because the *strength* of the relationship between the predictor and the criterion is the same.

4. *The correlation between Y and any linear transformation of X is numerically equal to r_{XY}.* Since Y' is a linear transformation of X, the correlation between X and Y is numerically equal to the correlation between Y and Y'. As we saw in Chapter 6, a linear transformation of a set of scores does *not* change the Z scores. (For example, adding a constant to a set of X scores does not change the Z_X scores.) This implies that a linear transformation of a set of X scores or a set of Y scores, or both, will not change the numerical value of the correlation between X and Y. And since Y' is a linear transformation of X, it also follows that the value of the correlation between X and Y will be equal to the correlation between the actual Y values and the predicted Y' values obtained from linear regression.

Thus, if the linear relationship between X and Y is poor (r_{XY} is near zero), predicted Y' scores obtained from linear regression will often be inaccurate and quite different from the actual Y scores ($r_{YY'}$ will be near zero). If the linear relationship between X and Y is strong (r_{XY} is numerically large), predicted Y' scores obtained from linear regression will often be accurate and consistent with the actual Y scores ($r_{YY'}$ will be large).

A Technical Note: Formulas for Predicting X From Y

Since the assignment of "X" and "Y" to the two variables in a linear regression problem is usually up to the researcher, you can usually assign the Y designation to the criterion and X to the predictor and use the formulas given previously. If you should then wish to make predictions in the other direction (because you want to entertain the opposite causal theory), you can simply *change the designations* (relabel X as Y and Y as X) and use the same formulas. Note that it *is* necessary to recompute a new linear regression equation if you wish to predict in the other direction (unless $r = \pm 1$).

For purposes of reference, we will note briefly the linear regression procedure for predicting scores on X from scores on Y. The equation is:

$$X' = b_{XY} Y + a_{XY}$$

where

$$X' = \text{predicted score on } X$$

$$b_{XY} = r_{XY} \frac{\sigma_X}{\sigma_Y} = \frac{N \sum XY - \sum X \sum Y}{N \sum Y^2 - (\sum Y)^2}$$

$$a_{XY} = \overline{X} - b_{XY}\overline{Y}$$

This procedure minimizes $\sum (X - X')^2$, the sum of squared errors when predicting X. While $r_{XY} = r_{YX}$, b_{XY} (the b value or slope for predicting from Y to X) does *not* equal b_{YX} (the b value or slope for predicting from X to Y), nor does $a_{XY} = a_{YX}$. Also, the regression line for predicting X is *not* the same as the regression line for predicting Y unless the correlation between X and Y is perfect (numerically equal to 1), $\sigma_X = \sigma_Y$, and $\overline{X} = \overline{Y}$.

Measuring Prediction Error: The Standard Error of Estimate

We have seen that when predicting scores on Y, the amount of squared error made for a given individual is equal to $(Y - Y')^2$. The *average* squared error for the entire sample can be obtained by summing the squared errors and dividing by N. Such a measure of error, the variance of errors, is in terms of squared units. As was the case with the standard deviation (Chapter 5), a measure in terms of actual score units can be obtained by taking the positive square root. This gives a useful measure of prediction error which is called the *standard error of estimate,* symbolized by $\sigma_{Y'}$:

$$\sigma_{Y'} = \sqrt{\frac{\sum (Y - Y')^2}{N}}$$

In practice, the standard error of estimate can more easily be obtained by the following formula, which can be proved to be equivalent to the one above by algebraic manipulation:

$$\sigma_{Y'} = \sigma_Y \sqrt{1 - r_{XY}^2}$$

Note that if $r = \pm 1$, there is no error ($\sigma_{Y'} = 0$); while if $r = 0$, the standard error of estimate reaches its maximum possible value σ_Y.

The standard error of estimate of Y may be thought of as the variability of Y about the regression line for a particular X value, averaged over all values of X. This is illustrated in Figure 12.10. In Figure 12.10A, the variability of the Y scores within each X value is relatively low, since the scores within each X value cluster closely together. The total variability of Y (for all X combined), which is given by σ_Y, is relatively large. Thus $\sigma_{Y'}$ is *small com-*

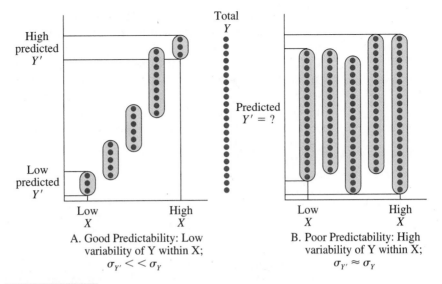

Relationship between the standard error of estimate ($\sigma_{Y'}$), the standard deviation of Y (σ_Y), and the accuracy of prediction

pared to σ_Y; the standard error of estimate is considerably smaller than the standard deviation of Y. This is good for purposes of prediction, since an individual with a low score on X is likely to have a low score on Y while an individual with a high score on X is likely to have a high score on Y. Thus, in this instance, Y takes on a narrow range of values for any given X.

Figure 12.10B, on the other hand, illustrates a situation in which prediction will be very poor. The average variability within each X value is about equal to the total variability of Y; that is, the standard error of estimate ($\sigma_{Y'}$) is *about equal to* the standard deviation of Y (σ_Y). In this instance, therefore, knowing an individual's X score will *not* permit you to make a good prediction as to what his Y score will be. Thus Figure 12.10 shows that prediction will be good to the extent that for any given X, Y shows little variability.

Note that in Figure 12.10A, X and Y are highly linearly correlated. In Figure 12.10B, the correlation is approximately zero. The standard error of estimate is also helpful in interpreting the meaning of a given numerical value of r, for it can be shown by algebraic manipulation that

$$r_{XY}^2 = 1 - \frac{\sigma_{Y'}^2}{\sigma_Y^2}$$

$$= \frac{\sigma_Y^2 - \sigma_{Y'}^2}{\sigma_Y^2}$$

That is, the *squared* correlation coefficient shows *by what proportion the reduction in the variability* (specifically the variance) *in predicting Y has* *been achieved by knowing X.* If this reduction has been great (as in Figure 12.10A), $\sigma^2_{Y'}$ is small compared to σ^2_Y, and r^2 is large. If, on the other hand, little reduction has been achieved (as in Figure 12.10B), $\sigma^2_{Y'}$ is about equal to σ^2_Y, and r^2 is about equal to zero. In the example involving GPA and SAT scores, r^2 is equal to $(.65)^2$ or .42. So in this example, *42% of the variance of Y is explained by X,* and a fairly good job has been done of identifying a (linear) relationship between two important variables.

The standard error of estimate computed from the formula given earlier is a descriptive statistic. (You may have inferred this from the fact that we used the Greek letter σ.) When you wish to draw inferences about the standard error of estimate in a population, you should instead use the following formula:

$$s_{Y'} = \sqrt{\frac{\sum (Y - Y')^2}{N - 2}}$$

The rationale is similar to the distinction between σ and s discussed in Chapter 5, except that here variability is measured about a line, not a point (the mean).

Summary

Correlation refers to the co-relationship between two variables. The Pearson *r* coefficient is a useful measure of *linear* correlation.

1. The Pearson correlation coefficient

Z score difference formula:

$$r_{XY} = 1 - \frac{1}{2} \frac{\sum (Z_X - Z_Y)^2}{N}$$

raw score definition formula:

$$r_{XY} = \frac{\sum (X - \bar{X})(Y - \bar{Y})}{N \sigma_X \sigma_Y}$$

Z score product formula:

$$r_{XY} = \frac{\sum Z_X Z_Y}{N}$$

raw score computing formula:

$$r_{XY} = \frac{N \sum XY - \sum X \sum Y}{\sqrt{[N \sum X^2 - (\sum X)^2]\,[N \sum Y^2 - (\sum Y)^2]}}$$

In the above formulas, N = number of *pairs.*

POINTS TO REMEMBER:

a. The numerical value of r indicates the *strength* of the relationship between the two variables, while the sign indicates the *direction* of the relationship.

b. $-1 \leq r \leq +1$.

c. r *cannot* be interpreted as a percent, but r^2 can.

d. r may be tested for statistical significance by referring the obtained value to the appropriate table with $df = N - 2$.

e. r may also be used as a descriptive statistic.

f. r detects only *linear* relationships.

g. Correlation does not imply causation.

h. The population correlation is symbolized by ρ.

2. *Linear regression*

The goal is to obtain the predicted score Y', given a score on X. The formula to use is:

$$Y' = b_{YX}X + a_{YX}$$

where

$$b_{YX} = r_{XY}\frac{\sigma_Y}{\sigma_X} = \frac{N\sum XY - \sum X \sum Y}{N\sum X^2 - (\sum X)^2}$$

$$a_{YX} = \overline{Y} - b_{YX}\overline{X}$$

The standard error of estimate for predicting Y from X, a measure of error in the prediction process, is equal to

$$\sigma_{Y'} = \sqrt{\frac{\sum (Y - Y')^2}{N}} = \sigma_Y\sqrt{1 - r_{XY}^2}$$

This measure is a descriptive statistic. If instead you wish to draw inferences about the standard error of estimate in a population, use the following formula:

$$s_{Y'} = \sqrt{\frac{\sum (Y - Y')^2}{N - 2}}$$

POINTS TO REMEMBER:

a. The above procedure minimizes the sum of squared errors in prediction, $\sum (Y - Y')^2$.

b. The closer r is to zero, the greater the errors in prediction. If $r = \pm 1$, $\sigma_{Y'} = 0$ (prediction is perfect). If $r = 0$, $\sigma_{Y'}$ equals its maximum possible value σ_Y (prediction is useless).

c. If r is not statistically significant, linear regression should *not* be used.

d. The regression line always passes through the point $\overline{X}, \overline{Y}$.

e. The above procedure is used only for purposes of predicting scores on Y (Y is the criterion and X is the predictor). Different formulas are needed to predict scores on X, and the simplest procedure in such cases is to change the designations (relabel X as Y and Y as X) so that the above procedure may be used.

Appendix: Equivalence of the Various Formulas for r

1.
$$1 - \frac{1}{2} \frac{\sum (Z_X - Z_Y)^2}{N} = \frac{\sum Z_X Z_Y}{N}$$

PROOF   Expanding the term in parentheses gives

$$r = 1 - \frac{1}{2} \frac{\sum (Z_X^2 - 2Z_X Z_Y + Z_Y^2)}{N} \; .$$

$$= 1 - \frac{1}{2} \frac{\sum Z_X^2 - 2 \sum Z_X Z_Y + \sum Z_Y^2}{N} \qquad \text{(Rules 1, 2, and 8, Chapter 1)}$$

$$= 1 - \frac{1}{2} \left(\frac{\sum Z_X^2}{N} - \frac{2 \sum Z_X Z_Y}{N} + \frac{\sum Z_Y^2}{N} \right)$$

It can readily be shown that the sum of squared Z scores divided by N is equal to 1:

$$\sigma^2 = \frac{\sum (Z - \overline{Z})^2}{N} \quad \text{(definition of } \sigma^2)$$

$$1 = \frac{\sum (Z - 0)^2}{N} \quad \text{(mean of } Z \text{ scores} = 0; \text{ variance of } Z \text{ scores} = 1)$$

$$1 = \frac{\sum Z^2}{N}$$

Thus,

$$r = 1 - \frac{1}{2} \left(1 - \frac{2 \sum Z_X Z_Y}{N} + 1 \right)$$

$$= 1 - 1 + \frac{\sum Z_X Z_Y}{N}$$

$$= \frac{\sum Z_X Z_Y}{N}$$

Incidentally, we know that the correlation of Z_X (or any other variable) with *itself* is $+1$, so $(\sum Z_X Z_X)/N = +1$. But this is the same as $(\sum Z_X^2)/N = 1$, which we have already seen.

2.
$$\frac{\sum Z_X Z_Y}{N} = \frac{\sum (X - \bar{X})(Y - \bar{Y})}{N \sigma_X \sigma_Y}$$

PROOF   By definition,

$$Z_X = \frac{X - \bar{X}}{\sigma_X}, \qquad Z_Y = \frac{Y - \bar{Y}}{\sigma_Y}$$

Therefore,

$$r = \frac{\sum \left(\dfrac{X - \bar{X}}{\sigma_X} \right) \left(\dfrac{Y - \bar{Y}}{\sigma_Y} \right)}{N}$$

$$= \frac{\sum (X - \bar{X})(Y - \bar{Y})}{N \sigma_X \sigma_Y} \qquad \text{(Rule 8, Chapter 1)}$$

3. Prove that

$$\frac{\sum (X - \bar{X})(Y - \bar{Y})}{N \sigma_X \sigma_Y}$$

and hence the other formulas in this appendix are equal to the raw score computing formula.

PROOF   We will deal with the numerator and denominator separately. Expanding the numerator gives

$$\sum (X - \bar{X})(Y - \bar{Y}) = \sum (XY - X\bar{Y} - \bar{X}Y + \bar{X}\bar{Y})$$
$$= \sum XY - \bar{Y} \sum X - \bar{X} \sum Y + N\bar{X}\bar{Y}$$

(Rules 1, 2, 5, and 8, Chapter 1)

Substituting $(\sum X)/N$ for $\bar{X}$ and $(\sum Y)/N$ for $\bar{Y}$ gives

$$\sum XY - \frac{(\sum Y)(\sum X)}{N} - \frac{(\sum X)(\sum Y)}{N} + \frac{N(\sum X)(\sum Y)}{(N)(N)}$$

$$= \sum XY - 2\frac{(\sum X)(\sum Y)}{N} + \frac{(\sum X)(\sum Y)}{N}$$

$$= \sum XY - \frac{\sum X \sum Y}{N}$$

Turning to the denominator, we have seen in Chapter 5 that

$$\sigma_X = \sqrt{\frac{1}{N}\left[\sum X^2 - \frac{(\sum X)^2}{N}\right]}$$

$$\sigma_Y = \sqrt{\frac{1}{N}\left[\sum Y^2 - \frac{(\sum Y)^2}{N}\right]}$$

Therefore,

$$N\sigma_X\,\sigma_Y = N\sqrt{\frac{1}{N}\left[\sum X^2 - \frac{(\sum X)^2}{N}\right]}\sqrt{\frac{1}{N}\left[\sum Y^2 - \frac{(\sum Y)^2}{N}\right]}$$

Since $[N(\sqrt{1/N})\,(\sqrt{1/N})]$ equals N/N or 1, these terms cancel out, leaving

$$\sqrt{\left[\sum X^2 - \frac{(\sum X)^2}{N}\right]\left[\sum Y^2 - \frac{(\sum Y)^2}{N}\right]}$$

If we now combine the numerator and denominator, we have

$$r = \frac{\sum XY - \dfrac{\sum X \sum Y}{N}}{\sqrt{\left[\sum X^2 - \dfrac{(\sum X)^2}{N}\right]\left[\sum Y^2 - \dfrac{(\sum Y)^2}{N}\right]}}$$

Finally, if we multiply this by $N/\sqrt{N^2}$, which is equal to 1, we have

$$r = \frac{N\sum XY - \sum X \sum Y}{\sqrt{[N\sum X^2 - (\sum X)^2]\,[N\sum Y^2 - (\sum Y)^2]}}$$

which is the raw score computing formula for r.

Chapter 13
Other Correlational Techniques

PREVIEW

The Relationship Between Ranked Variables: The Spearman Rank-Order Correlation Coefficient

How do ranks differ from continuous scores?

What are the procedures for computing the correlation between two variables when both consist of ranked data? Why do we use these procedures instead of the formula for the Pearson r?

How do we deal with tied ranks?

What are the procedures for testing the significance of the rank-order correlation coefficient? How do these procedures differ from the significance test for the Pearson r?

The Relationship Between One Dichotomous and One Continuous Variable

What is a dichotomous variable?

What are the procedures for computing the correlation between one continuous and one dichotomous variable (the point biserial correlation coefficient)? Why do we use these procedures instead of the formula for the Pearson r?

How is a point biserial correlation coefficient tested for statistical significance?

When we obtain a statistically significant value of t for the difference between two means, why is it desirable to convert this value to a point biserial correlation coefficient? How is this done?

When should you use the biserial correlation coefficient instead of the point biserial correlation coefficient?

The Relationship Between Two Dichotomous Variables

What correlation coefficient should be computed when both variables are genuinely dichotomous?

When should you use the tetrachoric correlation coefficient?

Summary

The Relationship Between Ranked Variables: The Spearman Rank-Order Correlation Coefficient

The Pearson correlation coefficient r is an appropriate statistic for describing the relationship between two variables when a straight-line model is appropriate, and when the data are continuous. For other kinds of data, however, other correlational techniques are necessary. Suppose that two psychologists *rank* a group of 10 patients in terms of degree of pathology. A rank of 1 indicates least pathological, while the person with a rank of 10 is most pathological. The results are shown in Table 13.1. Did the two psychologists rank the patients in a similar manner?

Ranks provide less information than do continuous scores. For example, in Table 13.1, it is evident that Observer X considered Robert as the least pathological and Joseph as the next lowest in pathology. With continuous scores, you could see at a glance whether these two patients are similar on this characteristic (as when Robert scores 72 on a personality test and Joseph scores 74) or different (Robert scores 72 and Joseph scores 95). With ranks, however, there is no indication of the *distance* between these two patients. Observer X might have felt that Robert was slightly less pathological than Joseph, *or* she might have felt that there was a considerable difference between them.

The statistic most frequently used to describe and to test hypotheses about the relationship between ranks is the Spearman rank-order correlation coefficient, symbolized as r_s.* It is computed as follows:

TABLE 13.1

Ten patients ranked for degree of pathology by two independent observers
(1 = least, 10 = most)

Patient's First Name	Ranks Given by Observer X	Ranks Given by Observer Y	D $(X - Y)$	D^2 $(X - Y)^2$
John	3	4	−1	1
Joseph	2	1	1	1
Charles	5	6	−1	1
William	9	7	2	4
Robert	1	3	−2	4
Bruce	10	10	0	0
Jennings	8	9	−1	1
Daniel	4	2	2	4
Richard	7	5	2	4
Raymond	6	8	−2	4
				$\sum D^2 = 24$

---

* The Greek letter rho, ρ, is also commonly used to designate the rank-order correlation coefficient. In order to avoid confusion between this statistic and the population parameter for the Pearson correlation coefficient, which is also symbolized as rho, we will use r_s for the Spearman coefficient.

$$r_s = 1 - \frac{6 \sum D^2}{N(N^2 - 1)}$$

where

D = **difference between a pair of ranks**
N = **number of pairs**

If you applied the formula for the Pearson product moment correlation co-efficient (Chapter 12) to rank-order data by treating the ranks as scores, and if there were no tied ranks, you would get exactly the same value as r_s. The formula for r_s is a shortcut for the formula for r; it takes advantage of certain relationships that hold when the N scores are the integers from 1 to N.

Like r, then, r_s varies between -1.0 and $+1.0$. A high positive value of r_s indicates a strong tendency for the paired ranks to be equal. Conversely, a high negative value of r_s indicates a strong tendency for the paired ranks to be opposite: patients ranked as most pathological by Observer X would be ranked at least pathological by Observer Y, an outcome that would undoubtedly produce some bewilderment on the part of the psychologists. A zero value would indicate no relationship between the two sets of ranks.

In our example, the calculation of $\sum D^2$ is shown in Table 13.1. First, the difference (D) between each pair of ranks is computed, where $D = X - Y$. Then, each difference is squared. Finally, the squared differences are summed. Then r_s is equal to

$$r_s = 1 - \frac{6(24)}{10(10^2 - 1)}$$

$$= 1 - \frac{144}{990}$$

$$= 1 - .15$$

$$= .85$$

Tied Ranks

In the preceding example, Observer X might have decided that Joseph and John were about equal in pathology and that both should share second place. This would be expressed numerically by *averaging* the two ranks in question, and assigning this average as the rank of each person. The ranks in question are 2 and 3, so the average of these ranks, or 2.5, is the rank given to both John and Joseph. The next patient would receive a rank of 4. What if three patients tied for the fourth rank? In this case, ranks 4, 5, and 6 would be averaged, and each of these three patients would receive a rank of 5. The next patient would be ranked 7th.

When there are tied ranks, the above formula will overestimate the absolute value of r_s. But unless there are many ties, and particularly long ties (that is, three- or more-way ties), this overestimate is likely to be trivial (less than .01 or .02).

Testing the Significance of the Rank-Order Correlation Coefficient

Although the formula for r_s and the formula for the Pearson r yield the same numerical result in the absence of tied ranks, the significance of the two coefficients *cannot* be tested in the same manner when N is very small (less than 10). To test the null hypothesis that the ranks are independent in the population from which the sample was drawn, you can use Table E in the Appendix. The minimum values of r_s needed for statistical significance are shown in the table for each N. (Note that when using this table, you need only refer to N—the number of pairs of ranks—rather than degrees of freedom.)

To use Table E for the problem involving the data in Table 13.1, you would look up the minimum value necessary for r_s to be statistically significant when $N = 10$. Using the two-tailed .05 criterion, this value is .648. Since the absolute value of the obtained r_s (.85) is greater than the tabled value, you reject H_0 and conclude that there is a nonzero relationship between the two sets of ranks in the population.

When N is greater than or equal to 10, the procedures for testing a Pearson r for statistical significance will give a very good approximation. That is, you can refer to the computed correlation to Table D in the Appendix with $N - 2$ degrees of freedom.

This test of significance is not, in general, as *powerful* as the test for the Pearson r. That is, it is *not* as likely to detect nonzero relationships. On the other hand, when one of a pair of variables is substantially skewed, transformation into ranks followed by the computation of r_s may well provide a better measure of their correlation than the Pearson r computed on the original data. A further discussion of this issue will be postponed until Chapter 18, where other tests based on ranks are presented under the heading of non-parametric and distribution-free methods.

The Relationship Between One Dichotomous and One Continuous Variable

The Point Biserial Correlation Coefficient

Suppose you are interested in the relationship between two variables, only one of which is continuous. The other is genuinely *dichotomous:* only two categories exist on the variable, such as male–female, presently married–presently unmarried, yes–no. You can compute the correlation between them by letting any number (say 0) stand for one of the dichotomized categories, using any other number (say 1) for the other category, and then using the usual formula for the Pearson r (Chapter 12). Thus the two numbers (for example, 0 and 1) are treated just as any other scores. It does not matter which two numbers you use because, as was pointed out in Chapter 12, the Pearson r *standardizes* each variable. Whatever numbers you use, the Z scores will not change in absolute value.

This correlation coefficient is called the *point biserial correlation coefficient*, symbolized as r_{pb}. While you can use the computational procedures discussed in Chapter 12 if you wish, there is an easier way to compute r_{pb}:

$$r_{pb} = \frac{\overline{Y}_1 - \overline{Y}_0}{\sigma_Y} \sqrt{pq}$$

Let the dichotomous variable (which equals 1 or 0) be symbolized by X and the continuous variable by Y. Then:

p = proportion scored 1 on the dichotomous variable (e.g., proportion of males)
q = proportion scored 0 on the dichotomous variable (e.g., proportion of females)
$\overline{Y}_1$ **= mean of the continuous scores for males**
$\overline{Y}_0$ **= mean of the continuous scores for females**
σ_Y **= standard deviation of the continuous scores for males and females combined**

As an illustration, consider the data in Table 13.2, where the goal is to determine the relationship between sex and arithmetic achievement scores for a random sample of high-school seniors. Arbitrarily, it is decided to score males as 1 on the dichotomous variable and females as 0, and r_{pb} is obtained as follows:

$$\sigma_Y = 9.15$$

$$p = \frac{8}{20} = \frac{2}{5} = .40$$

$$q = \frac{12}{20} = \frac{3}{5} = .60$$

$$\sqrt{pq} = \sqrt{(.40)(.60)} = \sqrt{.24} = .49$$

$$\overline{Y}_1 = \frac{10 + 15 + 15 + 20 + 5 + 10 + 10 + 5}{8} = 11.25$$

$$\overline{Y}_0 = \frac{30 + 20 + 25 + 20 + 25 + 30 + 5 + 10 + 20 + 30 + 35 + 10}{12}$$

$$= 21.67$$

$$r_{pb} = \left(\frac{11.25 - 21.67}{9.15} \right)(.49) = -.56$$

Thus, there is a high negative relationship between sex and achievement scores. Since males have been assigned a score of 1 and females a score of

TABLE 13.2

Scores on arithmetic achievement test for 20 high-school boys and girls

Sex (X)*	Achievement Scores (Y)
1	10
1	15
0	30
0	20
0	25
1	15
0	20
0	25
0	30
1	20
1	5
0	5
1	10
0	10
0	20
1	10
0	30
0	35
1	5
0	10

$$\sum Y = 350$$

$$(\sum Y)^2 = 122{,}500$$

$$\sum Y^2 = 7{,}800$$

$$\sigma_Y^2 = \frac{7{,}800 - \dfrac{122{,}500}{20}}{20} = 83.75$$

$$\sigma_Y = 9.15$$

* Male = 1, female = 0.

0, this means that there is a negative relationship between arithmetic achievement and being male or a positive relationship between arithmetic achievement and being female. If you had assigned females a score of 1 and males a score of 0, you would have obtained an r_{pb} of $+ .56$.

The point biserial correlation coefficient is tested for statistical significance in the same way as the Pearson r. Comparing the obtained value of $- .56$ to the critical value of .444 obtained from Table D in the Appendix for $N - 2$ or 18 df, you would decide to reject H_0 in favor of H_1 because the absolute value of the obtained r is greater than the tabled value. You would therefore conclude that a significant negative correlation exists between arithmetic achievement and being male in the population from which this sample of 20 paired observations was drawn.

Converting t Values for the Difference Between Two Means to r_{pb} Values

An alternative way of testing the significance of the data in Table 13.2 would be to form two groups, males and females, and apply the procedures of Chapter 11 to the data:

Males	Females
10	30
15	20
15	25
20	20
5	25
10	30
10	5
5	10
	20
	30
	35
	10

That is, you would compute

$$t = \frac{\overline{X}_{males} - \overline{X}_{females}}{\sqrt{S^2_{pooled}\left(\dfrac{1}{N_1} + \dfrac{1}{N_2}\right)}}$$

This procedure would yield exactly the same value of t you would get if you substituted the r_{pb} of $-.56$ in the formula for the t test of a Pearson correlation coefficient (Chapter 12). The point biserial correlation coefficient, however, has the advantage of providing an indication of the *size of the effect* as well as indicating statistical significance. In Chapter 12, we pointed out that a small Pearson r such as .08, even if statistically significant, would be almost useless for practical purposes. The size of the relationship between these two variables in the population, while likely to be greater than zero, is also likely to be extremely small.

This consideration is true of all procedures in inferential statistics. Statistical significance does *not* imply a *large* relationship or effect; it just indicates that the effect in the population is *unlikely to be zero*. A t value does *not* offer a solution to this problem, since it does not provide a ready measure of how large the relationship is (the effect size). So it is very desirable to convert values of t obtained from the procedures in Chapter 11 into a point biserial correlation coefficient, and to report *both* t (and statistical significance) and r_{pb}. It can be shown that this conversion can readily be accomplished by the following formula:

$$r_{pb} = \sqrt{\frac{t^2}{t^2 + df}}$$

When dealing with the t test for the difference between two means, df (as usual) is equal to $N_1 + N_2 - 2$.

As an illustration, consider once again the caffeine experiment discussed in Chapter 11. It was found that caffeine did have a positive effect on mathematics test scores; a statistically significant value of t was obtained. The question remains, however, as to *how strong* is the relationship between the presence or absence of that dosage of caffeine and test scores. Recalling that $t = 2.91$ and $df = (22 + 20 - 2) = 40$, we have

$$r_{pb} = \sqrt{\frac{(2.91)^2}{(2.91)^2 + 40}}$$

$$= \sqrt{\frac{8.47}{48.47}}$$

$$= \sqrt{.175}$$

$$= .42$$

Thus there is a moderately strong relationship between the presence or absence of caffeine and test scores. Since r_{pb}^2 can be interpreted in the same way as r^2, $(.42)^2$ or 17% of the variance in test scores is explained by whether or not caffeine is present. You therefore know that caffeine is a fairly important factor that accounts for differences among people on this test, though far from the only one—83% of the variance in test scores is still not accounted for. Assuming that caffeine has no known harmful side effects, a cautious recommendation that students take caffeine before a mathematics examination would be justified.

Suppose instead that, using a very large sample, the statistically significant t had yielded a value of r_{pb} of .10. Here, only $(.10)^2$ or 1% of the variance in test scores is explained by the presence or absence of caffeine. You are therefore warned that this experimental finding, while statistically significant, is probably not of much practical value. You certainly would *not* recommend that caffeine actually be used, for the probable effect on test scores would be too small to matter. (The possibility does exist, however, that such a finding might be of theoretical importance in understanding brain functioning.)

Unfortunately, it is very common to report values of t in the professional literature *without* also reporting the corresponding value of r_{pb}. Thus, there are many instances wherein researchers have exaggerated the importance of findings that are statistically significant, but that (unknown to them) are also so weak as to be almost useless in practice. We therefore suggest that you compute r_{pb} yourself when reading about t values in professional articles and books. You may well find that the researcher's enthusiastic commentary about a statistically significant result is unjustified.*

---

* For a more extensive discussion of converting t and other statistics into correlational-type indices, see J. Cohen, "Some statistical issues in psychological research," in B. B. Wolman ed., *Handbook of clinical psychology* (New York: McGraw-Hill, 1965).

The Biserial Correlation Coefficient

Suppose that a sample of 50 job applicants takes an ability test. All are then hired. Six months later, each employee's supervisor rates him or her as either a "success" or a "failure." (The supervisors do not feel able to give more precise ratings.) The question of interest is whether or not test scores, a continuous variable, are correlated with job success, a variable for which there are only two scores. Job success, however, is *not* a true dichotomy; there exists a continuum of job ability along which the employees differ. Unlike a genuine dichotomy such as male–female, the job success variable in this situation is actually a continuous variable on which only dichotomized scores are available.

In such cases, assuming that the variable which appears in crude dichotomous form is actually normally distributed and that it is this underlying variable whose correlation we want to estimate, it is recommended that a different correlation coefficient (called the *biserial* correlation coefficient) be computed instead of r_{pb}. A discussion of the biserial correlation coefficient, however, is beyond the scope of this book; the interested reader is referred to Guilford and Fruchter.*

The Relationship Between Two Dichotomous Variables

The Phi Coefficient

When you wish to determine the relationship between two variables and *both* of them are genuinely dichotomous, the appropriate correlation coefficient to compute is the *phi coefficient*. This coefficient is closely related to the chi-square statistic, so it will be discussed in the chapter dealing with chi square (Chapter 17).

The Tetrachoric Correlation Coefficient

In some situations, you may wish to determine the relationship between two variables where both are actually continuous, yet where only dichotomized scores are available on each one. As an illustration, suppose that 50 job applicants are interviewed. After each interview, the interviewer recommends either that the company hire or not hire the applicant. Regardless of the interviewer's rating, all 50 applicants are hired. Six months later, each employee is rated as either a "success" or a "failure." The company wishes to know the degree of relationship between the continuous, assumed normal variable that underlies the interviewer's recommendations and the similarly continuous normal variable underlying job success. Note that it is not the correlation between the observed dichotomies that is at issue — that would

---

* J. P. Guilford and B. Fruchter, *Fundamental statistics in psychology and education*, 6th ed. (New York: McGraw-Hill, 1978), 304–308.

be given by the phi coefficient—but rather the relationship between the underlying variables.

In situations such as this, it is recommended that a coefficient called the *tetrachoric correlation* be computed. Since a discussion of the tetrachoric correlation is beyond the scope of this book, we will again refer the interested reader to Guilford and Fruchter.*

Summary

1. The Spearman rank-order correlation coefficient

With ranked data, use the Spearman rank-order correlation coefficient:

$$r_s = 1 - \frac{6 \sum D^2}{N(N^2 - 1)}$$

where

$$D = \text{difference between a pair of ranks}$$
$$N = \text{number of pairs}$$

When $N < 10$, r_s must be referred to a different table than the Pearson r in order to determine statistical significance. For $N \geq 10$, the procedure used with the Pearson r gives a good approximation.

2. The point biserial correlation coefficient

With one continuous and one dichotomous variable, use the point biserial correlation coefficient:

$$r_{pb} = \frac{\overline{Y}_1 - \overline{Y}_0}{\sigma_Y} \sqrt{pq}$$

Let the dichotomous variable (equals 1 or 0) be symbolized by X, and let Y be the continuous variable. Then:

p = proportion scored 1 on the dichotomous variable
q = proportion scored 0 on the dichotomous variable
$\overline{Y}_1$ = mean score on Y for those scored 1 on the dichotomous variable
$\overline{Y}_0$ = mean score on Y for those scored 0 on the dichotomous variable
σ_Y = standard deviation of the Y scores

---

* Guilford and Fruchter, *op. cit.*, 311–316.

Alternatively, the correlation could be computed by using the formula for the Pearson r (Chapter 12). r_{pb} is tested for significance in the same way as the Pearson r.

3. *Converting significant values of t to r_{pb}*

When a statistically significant value of t in a test of the difference between two means is obtained, it is very desirable to convert it to r_{pb} so as to determine the *strength* of the relationship. This is readily done as follows:

$$r_{pb} = \sqrt{\frac{t^2}{t^2 + df}}$$

where

$$df = N_1 + N_2 - 2$$

4. *Chart summary of correlation coefficients*

The chart below summarizes the use of correlation coefficients discussed in this chapter. For example, if Variable B is continuous and Variable A is not a true dichotomy, the correct correlation coefficient to use to obtain a measure of relationship is biserial r.

		VARIABLE A	
	Continuous	Dichotomized (underlying distribution is actually normal)	Dichotomous (underlying distribution is actually two-valued)
Continuous	Pearson	biserial	point biserial
Dichotomized (underlying distribution is actually normal)	biserial	tetrachoric	
Dichotomous (underlying distribution is actually two-valued)	point biserial		phi

VARIABLE B

Chapter 14

Introduction to Power Analysis

PREVIEW

Introduction

What is the power of a statistical test? How is it related to a Type II error?

What are the practical consequences of making a Type II error?

Concepts of Power Analysis

How is the power of a statistical test related to the criterion of significance? To the size of the sample?

What is the population effect size, and how is it related to the power of a statistical test?

How large must the sample size be in order to use the procedures described in this chapter?

The Test of the Mean of a Single Population

What are the procedures for determining the power of statistical tests about the mean of a single population?

What are the procedures for determining the sample size required to have a specified power?

The Significance Test of the Proportion of a Single Population

What are the procedures for determining the power of statistical tests about the proportion of a single population?

What are the procedures for determining the sample size required to have a specified power?

The Significance Test of a Pearson r

What are the procedures for determining the power of statistical tests about the Pearson r?

Can these procedures also be used for rank-order, point-biserial, and phi coefficients?

What are the procedures for determining the sample size required to have a specified power?

(continued next page)

PREVIEW (continued)

Testing the Significance of the Difference Between Independent Means

What are the procedures for determining the power of statistical tests about the difference between two means, using independent samples?

What are the procedures for determining the sample size required to have a specified power?

Summary

Introduction

When you perform a statistical test of a null hypothesis, the probability of a Type I error (that is, rejecting H_0 when it is actually true) is equal to the significance criterion α (see Chapter 10). Since a small value of α is customarily chosen (for example, .05 or .01), you can be confident *when you reject H_0* that your decision is very likely to be correct.

In hypothesis testing, you also run a risk that even if H_0 is false, you may fail to reject it. That is, you may commit a Type II error. The probability of a Type II error is β. The complement of the probability of a Type II error, or $1 - \beta$, is the *probability of getting a significant result* and is called the *power* of the statistical test. Techniques for computing β or $1 - \beta$ were not discussed in Chapter 10.

Since the behavioral scientist who tests a null hypothesis almost certainly wants to reject it, he wants the power of the statistical test to be high rather than low. That is, he hopes that there is a good chance that the statistical test will indicate that he can reject H_0. Despite this, it is a strange fact that this topic has received little stress in the statistical textbooks used by behavioral scientists. The unfortunate result is that research is often done in which, unknown to the investigator, power is low (and, therefore, the probability of a Type II error is high), a false null hypothesis is not rejected (that is, a Type II error is actually made), and much time and effort are wasted. Worse, a promising line of research may be prematurely and mistakenly abandoned because the investigator does not know that he should have relatively little confidence concerning his failure to reject H_0.

This chapter will deal first with the concepts involved in power analysis. Then, methods will be presented for accomplishing the two major kinds of power analysis, which can be applied to the null hypothesis tests discussed thus far.

Concepts of Power Analysis

The material in this section, since it holds generally for a variety of statistical tests, is abstract and requires careful reading. In ensuing sections, these ideas will be incorporated into concrete procedures for power and sample size analysis of four major kinds of hypothesis tests.

There are four major factors involved in power analysis:

1. The significance criterion, α. This, of course, is the familiar criterion for rejecting the null hypothesis. It equals the probability of a Type I error, frequently .05. A little thought should convince you that the more stringent (the smaller) this criterion, other things being equal, the harder it is to reject H_0 and therefore the lower is the power. It also follows that, again other things including power being equal, a more stringent significance criterion requires a larger sample size for significance to be obtained.

2. The sample size, N. Whatever else the accuracy of a sample statistic may depend upon, it *always* depends on the size of the sample on which it has been determined. Thus, all the standard errors you have encountered in this book contain some function of N in the denominator. It follows that, other things being equal, error decreases and power increases as N increases.

3. The population "effect" size, γ (gamma). This is a new concept, and a very general one that applies to many statistical tests. When a null hypothesis about a population is false, it is false to some degree. (In other words, H_0 might be very wrong, or somewhat wrong, or only slightly wrong.) γ is a very general measure of the *degree* to which the null hypothesis is false, or how large the "effect" is in the population. Within the framework of hypothesis testing, γ can be looked upon as a *specific* value which is an alternative to H_0. Specific alternative hypotheses, in contrast to such universal alternative hypotheses as H_1: $\mu_1 - \mu_2 \neq 0$ or H_2: $\rho \neq 0$, are what make power analyses possible. We will consider γ in detail in subsequent sections.

Other things being equal, power increases as γ, the degree to which H_0 is false, increases. That is, you are more likely to reject a false H_0 if H_0 is very wrong. Similarly, other things including power being equal, the larger the γ, the smaller the N that is required for significance to be obtained.

4. Power, or 1 − β. The fourth parameter is power, the probability of rejecting H_0 for the given significance criterion. It is equal to the complement of the probability of a Type II error. That is, power = $1 - \beta$.

These four factors are mathematically related in such a way that any one of them is an exact function of the other three. We will deal with the two most useful ones:

1. Power determination. Given that a statistical test is performed with a specified α and N and that the population state of affairs is γ, the power of the statistical test can be determined.

2. N determination. Given that the population state of affairs is γ and a statistical test using α is to have some specified power (say .80), the necessary sample size, N, can be determined.

One final concept that will prove useful in statistical power analysis is δ (delta), where

$$\delta = \gamma \times f(N)$$

That is, δ is equal to γ times a function of N. Thus, δ combines the population effect size and the sample size into a single index. The table from which power is read (Table H in the Appendix) is entered using δ.

In the sections that follow, the general system described above for analyses of power determination and sample size determination will be implemented for four different statistical tests:

1. Tests of hypotheses about the mean of a single population (Chapter 10).
2. Tests of hypotheses about the proportion of a single population (Chapter 10).
3. Tests of the significance of a Pearson correlation coefficient (Chapter 12).
4. Tests of the difference between the means of two populations (Chapter 11).

The procedures to be described are approximate, and applicable for large samples (N at least 25 or 30). This is because the system uses the normal curve, which (as we have seen) is not only useful in its own right but is also a good approximation to the t distribution once N is that large.

The Test of the Mean of a Single Population

Power Determination

In Chapter 10, we described how to test the null hypothesis that the mean height of midwestern American men is 68 inches. The value of 68 in. was selected because it is the mean of U.S. adult males, and the issue is whether or not the mean of midwestern men differs from that height. Thus, the null hypothesis is: $H_0: \mu = \mu_0 = 68$.

The alternative hypothesis stated in Chapter 10 was merely that $H_1: \mu \neq 68$. However, power analysis is impossible unless a *specific* H_1 is stated. Assume that you suspect (or are interested in the possibility) that the population mean of midwestern men is an inch away from 68: H_1: $\mu = \mu_1 = 67$ or 69. (This form is used to indicate a two-tailed test. If a one-tailed test were desired such that midwestern men were predicted as taller, H_1 would be $\mu = \mu_1 = 69$.) Assume further that a two-tailed .05 decision rule is to be used, and that the sample size is to be $N = 100$. Finally, assume that you know or can estimate the population standard deviation (σ) of heights to be 3.1 in. The power determination question can then be formulated as follows: If we perform a test at $\alpha = .05$ of H_0: $\mu = 68$ using a random sample of $N = 100$, and in fact μ is 69 or 67, what is the probability that we will get a significant result and hence reject H_0?

The size of the effect postulated in the population, which is one inch in *raw score* terms, must be expressed as a γ value to accomplish the power analysis. For a test of the mean of a single population, γ is expressed essentially as a Z score:

$$\gamma = \frac{\mu_1 - \mu_0}{\sigma}$$

In the present example,

$$\gamma = \frac{69 - 68}{3.1} = \frac{1}{3.1} = .32$$

Note that this is *not* a statement about actual or prospective *sample* results. It expresses, as an alternative hypothesis, the *population* state of affairs. That is, μ_1 is postulated to be $.32\sigma$ away from μ_0, the value specified by H_0. Note also that the sign of γ is ignored, since the test is two-tailed.

Having obtained the measure of effect size, the next step is to obtain δ. In the previous section, we pointed out that $\delta = \gamma \times f(N)$. For the test of the mean of a single population, the specific function of N is $\sqrt{N}$, so that

$$\delta = \gamma\sqrt{N}$$

In the present example,

$$\delta = .32\sqrt{100} = 3.2$$

Entering Table H in the Appendix with $\delta = 3.2$ and $\alpha = .05$, the power is found to be .89. Thus, if the mean of the population of midwestern men is an inch away from 68, the probability of rejecting H_0 in this situation is .89 (or, the probability of a Type II error is .11). If the mean of the population of midwestern men is more than one inch away from 68, the power will be greater than .89. Conversely, if the population mean is less than one inch away from 68, the power will be less than .89.

It is important to understand that power analysis proceeds completely with population values. It does *not* utilize sample results, either actual or prospective. The above analysis could well take place before the data were gathered to determine what the power would be under the specified α, γ, and N. Or it could take place after an experiment was completed to determine the power the statistical test *had*, given α, γ, and N with no reference to the obtained data. If the power for a reasonably postulated γ were equal to .25 when the results of the experiment were not statistically significant, the nonsignificant result would be inconclusive, since the *a priori* probability of obtaining significance is so small (and the probability of a Type II error is so high). On the other hand, power of .90 with a nonsignificant result tends to suggest that the actual γ is not likely to be as large as postulated.

In order to compute γ in the above example, it was necessary to posit a specific value of μ_1 (69 or 67) and also of σ (3.1). This can sometimes be difficult when the unit is not a familiar one. (For example, extensive *a priori* data may not be available for a new test.) For situations like this, it is useful to specify conventional values corresponding to "small," "medium," and "large" values of γ, which although arbitrary are reasonable (in much the same way as the .05 decision rule). For the test of the mean of a single population, these are:

$$\text{small: } \gamma = .20$$

$$\text{medium: } \gamma = .50$$

$$\text{large: } \gamma = .80$$

If you are unable to posit specific values of μ_1 or σ, you can select the value of γ corresponding to how large you believe the effect size in the population to be. Do *not* use the above conventional values if you can specify γ values that are appropriate to the specific problem or field of research in which the statistical test occurs, for conventional values are only reasonable approximations. As you get to know a substantive research area, the need for reliance on these conventions should diminish.

To clarify the general logic of power analysis, let us reconsider the preceding problem in terms of the concepts introduced in Chapter 10. The null hypothesis states that $\mu_0 = 68$ and the standard error of the mean is equal to .31 ($\sigma/\sqrt{N}$ is equal to 3.1/10), so the sampling distribution of means postulated by H_0 can be illustrated by the graph shown at the *top* in Figure 14.1. Assume that the true value of μ is equal to 69, for which the sampling distribution of means is as shown by the graph at the *bottom* in Figure 14.1. How likely is it that you will retain the incorrect μ_0 of 68?

Since we are using the normal curve as an approximation to the t distribution, the critical value (z_0) is equal to ±1.96. If a sample mean is obtained that (when expressed as a z value) is numerically less than 1.96, the incorrect H_0 will be retained. This z_0 value can readily be converted to a critical

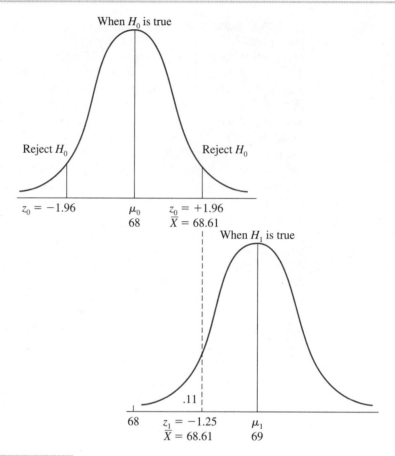

FIGURE 14.1
Sampling distribution of means ($N = 100$) assuming $\mu_0 = 68$ and $\mu_1 = 69$

value of the sample mean:

$$z_0 = \frac{\overline{X} - \mu_0}{s_{\overline{x}}}$$

$$1.96 = \frac{\overline{X} - 68}{.31}$$

$$\overline{X} = (1.96)(.31) + 68$$

$$\overline{X} = 68.61$$

So the incorrect H_0 will be rejected if the sample mean is greater than 68.61. To see how likely this is to occur when the population mean is actually equal to 69, consider only the bottom graph in Figure 14.1 and convert the mean value of 68.61 to a z value:

$$z_1 = \frac{\overline{X} - \mu_1}{s_{\overline{x}}}$$

$$z_1 = \frac{68.61 - 69}{.31} = -1.25$$

The percent of the area under the normal curve to the right of this z value can be found by using Table B: at $z = -1.25$, it is about 89% ($= 39.44\% + 50.00\%$). Therefore, when $\mu = 69$, about 89% of the sample means which one can obtain would be greater than 68.61 and thus significant. So the probability of rejecting H_0 when $\mu = 69$, and hence the power of this test, is .89. Since the test is two-tailed and the analysis symmetrical, the same argument would lead to the same conclusion if the true $\mu = 67$, one inch *below* $H_0 = 68$. Thus we can conclude that under these conditions, if $\mu = 69$ *or* 67, the probability of rejecting H_0 ($\mu_0 = 68$) is .89.

If the null hypothesis is so wrong that there is virtually no chance of a Type II error (that is, if the effect size is extremely large), power will be virtually perfect (approaching 1.0). This does *not* indicate that the research is necessarily good, for the researcher may be testing a proposition so obvious as to be useless. (For example, his null hypothesis might be that the mean IQ of college graduates is 68!) If, on the other hand, the null hypothesis is only slightly wrong (the researcher is studying a small-sized effect), a large sample size will be needed in order for the research to have acceptable power.

Sample Size Determination

We now turn to the other major kind of statistical power analysis, sample size determination. Here, α, γ, and the desired power are specified, and you wish to determine the necessary sample size. This form of analysis is particularly useful in experimental planning, since it provides the only rational method for making the crucially important decision about the size of N.

First, we must consider the concept of "desired power." Your first inclination might be to set power very high, such as .99 or .999. But, as your intuition might suggest, the quest for near certainty is likely to result in the requirement of a very large N, usually far beyond your resources. This is similar to the drawback of choosing a very small significance criterion, such as .0001: while a Type I error is very unlikely with such a criterion, you are also very unlikely to obtain statistical significance unless the effect size and/or sample size is unrealistically large. Therefore, just as it is prudent to seek less than certainty in minimizing Type I errors by being content with a significance criterion such as .05 or .01, a prudent power value is also in order insofar as Type II errors are concerned.

Although a researcher is of course free to set any power value that makes sense to her, we suggest the value .80 (which makes the probability of a Type II error equal to .20) when a conventional standard is desired. Why is this suggested value larger than the customary .05 value for the probability of a Type I error? In most instances in science, it is seen as less desirable to mistakenly reject a true H_0 (which leads to false positive claims) than to mistakenly fail to reject H_0 (which leads only to the failure to find something and no claim at all: "the evidence is insufficient to warrant the conclusion that . . ."). Besides, if desired power were conventionally set at .95 (making $\beta = .05$), most studies would demand larger samples than most investigators could muster. We repeat, however, that the .80 value should be used only as a general standard, with investigators quite free to make their own decision.

Let us return to the mean height of midwestern men, but now change the task. Instead of assuming that N is to be 100, suppose you are now interested in the following question: "If I test at $\alpha = .05$ the null hypothesis that $\mu = 68$ when in fact $\mu = 69$ or 67 (and $\sigma = 3.1$), how large must N be for me to have a .80 probability of rejecting H_0 (that is, to have power = .80)?" To answer this question, the first step is to obtain two values: γ, determined by the same methods as in the preceding section and equal to $(\mu_1 - \mu_0)/\sigma$ or .32, and the desired power, specified as .80. Next, δ is obtained by entering Table I in the Appendix in the row for desired power = .80 and the column for α (two-tailed) = .05; δ is found to be 2.80. Finally, for this test of the mean of a single population, N is found as follows:

$$N = \left(\frac{\delta}{\gamma} \right)^2$$

In the present example,

$$N = \left(\frac{2.80}{.32} \right)^2$$

$$= (8.75)^2$$

$$= 77$$

To have power = .80 in the present situation, the sample must have 77 cases in it. Note that this is consistent with the previous result where we saw that, other things (γ, α) equal, $N = 100$ resulted in power = .89.

To illustrate the consequences of demanding very high power, consider what happens in this problem if desired power is set at .999. From Table I in the Appendix, $\delta = 5.05$. Substituting,

$$N = \left(\frac{\delta}{\gamma} \right)^2 = \left(\frac{5.05}{.32} \right)^2 = (15.78)^2 = 249$$

In this problem, to go from .80 to .999 power requires increasing N from 77 to 249. This is more than many researchers can manage. Of course, if data are easily obtained or if the cost of making a Type II error is great, no objection can be raised about such a "maximum" power demand.

The Significance Test of the Proportion of a Single Population

Power Determination

The structure of the system remains the same as before; only the details need to be adjusted. The null hypothesis in question is that the population proportion, π, is equal to some specified value. That is, $H_0: \pi = \pi_0$. H_0 is tested against some specific alternative, $H_1: \pi = \pi_1$.

Let us return to the worried politician in Chapter 10, who wants to forecast the results of an upcoming two-person election by obtaining a random sample of $N = 400$ voters and testing the null hypothesis that the proportion favoring him is $\pi_0 = .50$. He thinks that he is separated from his opponent by about .08. That is, he expects the vote to be .54 to .46 or .46 to .54. (His expectations are stated in both directions because a two-tailed significance test is intended.) The question can be summarized as follows: If a statistical test is performed at $\alpha = .05$ of $H_0: \pi = .50$ against the specific alternative $H_1: \pi = .54$ (or .46) with $N = 400$, what is the power of this statistical test?

For the test of a proportion from a single population, the effect size, γ, and δ are defined as follows:

$$\gamma = \frac{\pi_1 - \pi_0}{\sqrt{\pi_0 (1 - \pi_0)}}$$

$$\delta = \gamma \sqrt{N}$$

For the data of this problem,

$$\gamma = \frac{.54 - .50}{\sqrt{.50 (1 - .50)}}$$

$$= \frac{.04}{.50}$$

$$= .08$$

$$\delta = .08 \sqrt{400}$$

$$= 1.60$$

Entering Table H with $\delta = 1.60$ and $\alpha = .05$, the power is found to be .36. So our worried politician has something else to worry about: as he has planned the study, he has only about one chance in three of coming to a positive conclusion (rejecting the null hypothesis that $\pi = .50$) if the race is as close as he thinks (H_1: $\pi = .54$ or $.46$). If the poll had already been conducted as described, and the results were not statistically significant, he should consider the results inconclusive. Even if π is as far from .50 as .54 or .46, the probability of a Type II error (β) is .64.

Sample Size Determination

Let us now invert the problem to one of determining N, given the same $\gamma = .08$, $\alpha = .05$ and specifying the desired power as .80. The formula for N for the test of a proportion from a single population, like the above formula for δ, is the same as in the case of the mean of a single population:

$$N = \left(\frac{\delta}{\gamma}\right)^2$$

The value of δ for the joint specification of power $= .80$, α (two-tailed) $= .05$ is found from Table I to be 2.80. Therefore,

$$N = \left(\frac{2.80}{.08}\right)^2$$

$$= (35)^2$$

$$= 1,225$$

This sample size is considerably larger than the $N = 400$ that yielded power of .36 under these conditions. It is hardly a coincidence that in surveys conducted both for political polling and market and advertising research, sample sizes typically run about 1,500.

If conventional values for effect size for this test are needed, the following γ values can be used: small, .10; medium, .30; large, .50.

The Significance Test of a Pearson *r*

Power Determination

The term Pearson *r* is intended to be used inclusively for *all* such *r*s; it includes the rank-order, point-biserial, and phi coefficients, as well as those computed from continuous interval scales. Power analysis of significance tests of the Pearson *r* is quite simple. Recall from Chapter 12 that for such tests, the null hypothesis is: H_0: $\rho = 0$. For the purpose of power analysis, the alternative hypothesis is that the population value is some *specific* value other than zero: H_1: $\rho = \rho_1$.

For example, you might be testing an *r* for significance when expecting that the population value is .30. In this case, H_1: $\rho = .30$. (Again, for two-tailed tests, the value is taken as either $+.30$ or $-.30$.) The γ value for this test requires no formula; it is simply the value of ρ specified by H_1 (.30 in this example). To find δ, the appropriate function of N to be combined with γ is $\sqrt{N-1}$:

$$\delta = \gamma \sqrt{N-1} = \rho_1 \sqrt{N-1}$$

Consider a study in which it is to be determined whether there is a linear relationship between a questionnaire measure of introversion and some physiological measure. A reasonably random sample of 50 Psychology 1 students at a large university is used for a two-tailed statistical test with $\alpha =$ .01. The researcher expects the population ρ to be .30; thus $\rho_1 = .30$ while H_0: $\rho = 0$. What is the power of the test?

We can go directly to

$$\delta = .30 \sqrt{50 - 1}$$

$$= 2.10$$

Entering Table H for $\delta = 2.10$ and $\alpha = .01$, power is found to be .32. A one chance in three of finding significance may strike the researcher as hardly worth the effort. He may then reconsider the stringency of his significance criterion, and check $\alpha = .05$; the power of .56 with this more lenient criterion may still not satisfy him. He might then plan to increase his sample size (as shown in the next section).

If the investigator has difficulty in formulating an alternative-hypothetical value for ρ_1, the following conventional values are offered: small, .10; medium, .30; large, .50. The value of .50 may not seem "large," but over most of the range of behavioral science where correlation is used, correlations between different variables do not often get much larger than that.

Sample Size Determination

Returning to the introversion study above, we can ask what N is necessary for a test at α (two-tailed) $= .05$ and assuming $\gamma = \rho_1 = .30$ in order to have power of (let us say) .75. The value of N is just one more than for the other two one-sample tests described above:

$$N = \left(\frac{\delta}{\gamma}\right)^2 + 1 = \left(\frac{\delta}{\rho_1}\right)^2 + 1$$

δ is found from Table I for power $= .75$, $\alpha = .05$, to be 2.63, so

$$N = \left(\frac{2.63}{.30}\right)^2 + 1$$

$$= (8.77)^2 + 1$$

$$= 78$$

To have a .75 chance (that is, three-to-one odds) of finding r to be significant if $\rho_1 = +.30$ (or $-.30$), he needs a sample of 78 cases.

Testing the Significance of the Difference Between Independent Means

The final significance test whose power analysis we consider is the test of the difference between the means of two independently drawn random samples. This is probably the most frequently performed test in the behavioral sciences.

As stated in Chapter 11, the null hypothesis most frequently tested is H_0: $\mu_1 - \mu_2 = 0$ (often written as $\mu_1 = \mu_2$). For power analysis, a *specific* alternative hypothesis is needed. We will write this as H_1: $\mu_1 - \mu_2 = \theta$ (theta), where θ is the difference between the means expressed in *raw* units. To obtain γ, the standard measure of effect size, this difference must be standardized. This is done using the standard deviation of the population, σ (which is a single value since we assume that $\sigma_1 = \sigma_2$ for the two populations). Thus the value of γ in this instance is conceptually the same as for the test of a single population mean, namely

$$\gamma = \frac{\text{alternative-hypothetical } \mu_1 - \mu_2}{\sigma} = \frac{\theta}{\sigma}$$

This can be looked upon as the difference between Z score means of the two populations, or, equivalently, the difference in means expressed in units of σ. Again, the direction of the difference (the sign of θ) is ignored in two-tailed tests.

This device of "standardizing" the difference between two means is generally useful, and not only for purposes of power analysis. Frequently in behavioral science, there is no sure sense of how large a unit of raw score is. How large *is* a point? Well, they come in different sizes: An IQ point comes about 15 to the standard deviation (σ_{IQ}) while an SAT point is much smaller, coming 100 to the standard deviation (σ_{SAT}). By always using σ as the unit of measurement, we achieve comparability from one measure to another, as was the case with Z scores.

This device helps us with regard to other related problems. One occurs whenever we have not had much experience with a measure (a new test, for example) and we have little if any basis for estimating the population σ, which the above formula for γ requires. Paradoxically enough, we can use this σ as our unit, despite the fact that it is unknown, by thinking directly in terms of γ. Thus a γ of .25 indicates a difference between two population means (whose exact value we do not know) equal to $.25\sigma$ (whose exact value we also do not know). It is as if our ignorance cancels out, and in a most useful sense $\gamma = .25$ is always the same size difference whether we are talking about IQ, height, socioeconomic status, or a brand-new measure of "oedipal intensity."

How large *is* a γ of .25, or any other? Here, as before, it is possible (and sometimes necessary) to appeal to some conventions, and they are the same as in the test of the mean of a single population: small, .20; medium, .50; large, .80. In this framework, a γ of .25 would be characterized as a "smallish" difference. Again we point out that when an investigator has a firmer basis for formulating γ, these conventions should not be used. Frequently, however, they will come in handy.

Power Determination

For the comparison of two means, the value of δ is equal to:

$$\delta = \gamma \sqrt{\frac{N}{2}}$$

where

N = size of *each* of the two samples (thus 2N cases are needed in all)

As an illustration, consider the caffeine experiment in Chapter 11 (slightly revised). Given two groups, each with $N = 40$ cases, one of which is given a dose of caffeine and the other a placebo, we will test at $\alpha = .05$ (two-tailed) the difference between their means on a mathematics test following the treatment. If there is in the populations a mean difference of medium size, operationally defined as $\gamma = .50$, what is the power of the test? With γ given on the basis of convention, there is no need to compute it, and hence no need to estimate μ_1, μ_2, or σ. Thus, all we need is

$$\delta = .50 \sqrt{\frac{40}{2}}$$

$$= 2.24$$

Entering Table H in the column for $\alpha = .05$, we find for $\delta = 2.2$ that power = .59 and for $\delta = 2.3$ that power = .63. Interpolating between these two values to obtain the power corresponding to $\delta = 2.24$ yields power of $.59 + .4(.63 - .59) = .61$. Thus, for medium differences (as defined here), samples of 40 cases each have only a .61 probability of rejecting H_0 for a two-tailed test using $\alpha = .05$.

Ordinarily, in research where variables are to be manipulated such as the caffeine experiment described above, it is possible to arrange the experiment so that the total available pool of subjects is divided equally among the groups. This division is optimal. But it is not always possible or desirable to have equal sample sizes. For example, it would be a mistake to reduce the larger of two available samples to make it equal to the smaller. Instead, to determine power when $N_1 \neq N_2$, compute the so-called *harmonic mean* of the two *N*s:

$$\text{harmonic mean} = \frac{2N_1N_2}{N_1 + N_2}$$

Then, use this obtained value for *N* in the formula for δ. As an example, consider a psychiatric hospital study where there are resources to place a random sample of 50 newly admitted schizophrenic patients into a special experimental treatment program. After one month, their progress is compared with a control group of 150 cases receiving regular treatment for the same period. It is planned to rate each patient on an improvement scale 30 days following admission and to test for the significance of the difference in means for rated improvement, using a two-tailed test where α = .05. If it is anticipated that γ = .40 (between "small" and "medium"), what is the power of the test?

Since $N_1 \neq N_2$, compute

$$\frac{2N_1N_2}{N_1 + N_2} = \frac{2\,(50)(150)}{50 + 150} = 75$$

Then enter this value as *N* in the formula for δ:

$$\delta = .40 \sqrt{\frac{75}{2}}$$

$$= 2.45$$

Interpolating between δ = 2.4 and δ = 2.5 in Table H in the α = .05 column yields power = .69.

Note that a total of $N_1 = N_2 = 200$ cases will be studied. But since $N_1 \neq N_2$, the resulting power is the same as if two samples of 75 cases each, or 150 cases, were to be studied. There is a smaller effective *N*, and hence less power. This demonstrates the nonoptimality of unequal sample sizes. Were it possible to divide the total of 200 cases into two equal samples of 100, δ would equal .40 $\sqrt{100/2}$ or 2.83, and the resulting power at α = .05 obtained from Table H by interpolation is .81.

Sample Size Determination

The sample size needed for *each* independent sample when means are to be compared can be found for specified values of α, γ, and the desired power, as was shown for the other statistical tests in this chapter. The combination of α and power is expressed as δ by looking in Table I, and the values for δ and γ are substituted in the equation appropriate for this statistical test, which is

$$N = 2 \left(\frac{\delta}{\gamma} \right)^2$$

For example, it was found in the caffeine experiment that for $\alpha = .05$ and $\gamma = .50$, the plan to use samples of $N = 40$ each resulted in power $= .61$. If power $= .90$ is desired, what N per sample is needed?

For a two-tailed test where $\alpha = .05$ and power $= .90$, Table I yields $\delta = 3.24$. Substituting,

$$N = 2 \left(\frac{3.24}{.50} \right)^2$$

$$= 84 \text{ cases in each group, or 168 cases in all}$$

Again we see that large desired power values demand large sample sizes. Were the more modest power of .80 specified, δ would equal 2.80 (Table I) and N would equal $2(2.80/.50)^2$ or 63, a materially smaller number.

This chapter, as its title indicates, merely introduces the concepts of power analysis and presents some simple approximate procedures for performing it in connection with four statistical tests. For a more thorough but still essentially nontechnical presentation, which covers all the major statistical tests with many tables that minimize the need for computation, see Cohen.[*]

Summary

1. The meaning of power

Power is the *probability of getting a significant result* in a statistical test. Power is equal to $1 - \beta$, where $\beta =$ the probability of a Type II error.

If power is not known, a researcher may waste a great deal of time by conducting an experiment that has little chance to produce significance even if H_0 if false. Worse, a promising line of research may be abandoned because the researcher does not know that he should have relatively little confidence concerning his failure to reject H_0 (that is, the probability of a Type II error is large).

2. The four major parameters of power analysis

1. *The significance criterion,* α. A Type II error is more likely as α gets smaller because you fail to reject H_0 more often. Thus, power decreases as α decreases.
2. *The sample size, N.* Larger samples yield better estimates of population parameters and make it more likely that you will reject H_0 when it is correct to do so. Thus, power increases as N increases.

---

[*] J. Cohen, *Statistical power analysis for the behavioral sciences,* 2nd ed. (Hillsdale, NJ: Erlbaum, 1977).

3. *The population "effect" size,* γ *(gamma).* You are less likely to fail to reject a false H_0 if H_0 is "very wrong" — that is, if there is a large effect size in the population. Thus, power increases as γ increases.

4. *Power,* $1 - \beta$.

Any one of these four parameters is an exact mathematical function of the other three.

3. The general procedure for the two most important kinds of power analysis

1. Power determination
   a. Compute or posit the value of γ.
   b. Compute δ (delta), which combines N and γ.
   c. Obtain power from the appropriate table.

2. Sample size determination
   a. Specify desired power. (If a conventional value is needed, use .80.)
   b. Compute or posit the value of γ.
   c. Obtain δ from the appropriate table.
   d. Compute N.

If you are unable to specify the population values needed to compute γ, use the appropriate conventional value of γ for the statistical test in question.

4. The specific formulas for γ, δ, and N depend on the statistical test that is being performed.

1. The test of the mean of a single population

   a. $\gamma = \dfrac{\mu_1 - \mu_0}{\sigma}$

   where

   > μ_0 = value of μ specified by H_0
   > μ_1 = value of μ specified by H_1
   >   (a *specific* H_1 is always necessary in power analysis)
   > σ = population standard deviation

   b. $\delta = \gamma \sqrt{N}$

   c. $N = \left(\dfrac{\delta}{\gamma}\right)^2$

   d. Conventional values of γ: "small," .20; "medium," .50; "large," .80

2. The test of the proportion of a single population

    a. $\gamma = \dfrac{\pi_1 - \pi_0}{\sqrt{\pi_0 (1 - \pi_0)}}$

    where

$$\pi_0 = \text{value of } \pi \text{ specified by } H_0$$
$$\pi_1 = \text{value of } \pi \text{ specified by } H_1$$

    b. $\delta = \gamma \sqrt{N}$

    c. $N = \left(\dfrac{\delta}{\gamma}\right)^2$

    d. Conventional values of γ: "small," .10; "medium," .30; "large," .50

3. The significance test of a Pearson r (includes Pearson r with continuous interval scales, r_s, r_{pb}, ϕ)

    a. $\gamma = \rho_1$, **the value specified by H_1**

    b. $\delta = \gamma \sqrt{N - 1}$

    c. $N = \left(\dfrac{\delta}{\gamma}\right)^2 + 1$

      $= \left(\dfrac{\delta}{\rho_1}\right)^2 + 1$

    d. Conventional values of γ: "small," .10; "medium," .30; "large," .50

4. The significance test of the difference between independent means

    a. $\gamma = \dfrac{\theta}{\sigma}$

    where

$$\theta = \text{value of } \mu_1 - \mu_2 \text{ specified by } H_1$$
$$\sigma = \text{population standard deviation (a single}$$
$$\text{value because it is assumed that } \sigma_1 = \sigma_2)$$

    b. $\delta = \gamma \sqrt{\dfrac{N}{2}}$

where

N is the size of *each* sample
(thus 2N cases are used in the experiment)

If the sample sizes are unequal, use

$$N = \frac{2\, N_1 N_2}{N_1 + N_2}$$

c. $N = 2\left(\dfrac{\delta}{\gamma}\right)^2$

where

N is the size of *each* sample
(so 2N cases will be needed in all)

d. Conventional values of γ: "small," .20; "medium," .50; "large," .80

Note: The above procedures are approximate, and are applicable for large samples (N at least 25 or 30).

Chapter 15

One-Way Analysis of Variance

PREVIEW

Introduction

We wish to draw inferences about the differences between *more* than two population means. Why is it a poor idea to perform numerous *t* tests between the various pairs of means?

What is the experimentwise error rate?

In what way is one-way analysis of variance similar to the *t* test for the difference between two means? How does it differ?

The General Logic of ANOVA

We wish to draw inferences about population means. Why then does the statistical analysis use variances?

What is the within-group (or error) variance estimate? What is the between-group variance estimate?

What is the *F* ratio?

Computational Procedures

What are the procedures for drawing inferences about any number of population means by computing a one-way analysis of variance? What are sums of squares? Mean squares? The *F* ratio?

What is the correct statistical model to use in this situation?

How do we test the *F* ratio for statistical significance?

One-Way ANOVA With Unequal Sample Sizes

What are the procedures for computing a one-way analysis of variance when sample sizes are *not* equal?

Multiple Comparisons: The Protected *t* Test

When the *F* ratio is statistically significant, what important additional information is provided by multiple comparisons?

What are the procedures for computing the protected *t* test?

When should the protected *t* test *not* be used?

(continued next page)

PREVIEW (continued)

How are confidence intervals used in the protected *t* test? What is their
advantage?

Some Comments on the Use of ANOVA

What assumptions underlie the use of one-way analysis of variance?

Why is it desirable to convert a statistically significant value of *F* into a measure
that shows the strength of the relationship between the independent and
dependent variable? How is this done?

Summary

**Appendix: Proof That the Total Sum of Squares Is Equal to
the Sum of the Between-Group and the Within-Group Sum
of Squares**

Introduction

In Chapter 11, techniques were given for testing the significance of the difference between two means. These techniques enable you to determine the
effect of a single independent variable (for example, the presence or absence
of a particular dosage of caffeine) on the mean of a dependent variable
(mathematics test scores) when there are *two* samples of interest, such as an
experimental group and a control group.

Suppose that you would like to determine whether *different dosages* of
caffeine affect performance on a 20-item English test. As before, there is one
independent variable (amount of caffeine), but this time you wish to include
the following *five* samples:

Sample 1. very large dose of caffeine

Sample 2. large dose of caffeine

Sample 3. moderate dose of caffeine

Sample 4. small dose of caffeine

Sample 5. placebo (no caffeine)

As usual, you would like the probability of a Type I error in this experiment to be .05 or less. It would *not* be correct to perform ten separate *t* tests
for the difference between two means (that is, first test $H_0: \mu_1 = \mu_2$; then
test $H_0: \mu_1 = \mu_3$; and so on), and then compare each obtained *t* value to the
appropriate critical *t* value for $\alpha = .05$.

The more statistical tests you perform, the more likely it is that some will
be statistically significant purely by chance. That is, when $\alpha = .05$, the prob-

ability that one t test will yield statistical significance when H_0 actually is true is .05. This is the probability of committing a Type I error. If you run 20 t tests when H_0 is always true, an average of one of them $(.05 \times 20 = 1.0)$ will be statistically significant just on the basis of chance, so it is very likely that you will commit a Type I error somewhere along the line. Similarly, if you run ten separate t tests in order to test the null hypothesis that different dosages of caffeine do not affect English test scores, the probability that you will commit at least one Type I error is clearly *greater* than the desired .05. (It can be shown that this probability is approximately .29.) The greater the number of sample means among which you perform pairwise significance tests, the more likely you are to make at least one Type I error. In statistical terminology, the rate of occurrence of *any* Type I errors over a *series* of individual statistical tests is called the *experimentwise* error rate.

Fortunately, there is a procedure for testing differences among three or more means for statistical significance which overcomes this difficulty. This procedure is the *analysis of variance,* abbreviated as ANOVA. ANOVA can also be used with just two samples, in which case it yields identical results to the procedures given in Chapter 11. The null hypothesis tested by ANOVA is that the means of the populations from which the samples were randomly drawn are all equal. For example, the null hypothesis in the caffeine experiment is

$$H_0: \mu_1 = \mu_2 = \mu_3 = \mu_4 = \mu_5$$

The alternative hypothesis merely states that H_0 taken as a whole is *not* true. There are many ways in which it may be false: it may be that $\mu_1 \neq \mu_2$, or that $\mu_3 \neq \mu_5$, or both, or that all five population means are unequal, or only two are equal, and so forth. The rejection of the ANOVA H_0 tells us only that *some* inequality exists. Later in this chapter, we will describe a separate procedure for testing pairwise mean differences following an ANOVA. This procedure enables us to specify the meaning of rejecting the ANOVA H_0.

The General Logic of ANOVA

Since it may seem paradoxical to test a null hypothesis about *means* by testing *variances,* the general logic of ANOVA will be discussed before proceeding to the computational procedures. The ANOVA procedure is based on a mathematical proof that the sample data can be made to yield two independent estimates of the population variance:

1. Within-group (or "error") variance estimate.   This estimate is based on how different each of the scores in a given sample (or *group*) is from other scores in the same group.

2. Between-group variance estimate.   This estimate is based on how different the *means* of the various samples (or *groups*) are from one another.

If the samples all come from the same normally distributed population (or from normally distributed populations with equal means and variances), it can be proved mathematically that the between-group variance estimate and the within-group variance estimate will be about equal to each other, and equal to the population variance (σ^2). The larger the between-group variance estimate is in comparison to the within-group variance estimate, the more likely it is that the samples do *not* come from populations with equal means.

As an illustration, consider the hypothetical data in Table 15.1A. There are five groups; Group 1 = largest dose of caffeine, and Group 5 = placebo. Each group contains five people, so there are 25 subjects in all, and the entries in the table represent the score of each person on the English test. There is some between-group variation and some within-group variation. The means of the five groups may differ because caffeine does have an effect on test scores, or merely because they are affected by the presence of sampling error (that is, the accident of which cases happened to be in each sample), or for both reasons. Whether to retain or reject the null hypothesis that the samples all come from the same population is decided by applying ANOVA procedures to the data. Inspection of Table 15.1A should suggest

TABLE 15.1

Two possible outcomes of the caffeine experiment (hypothetical data)

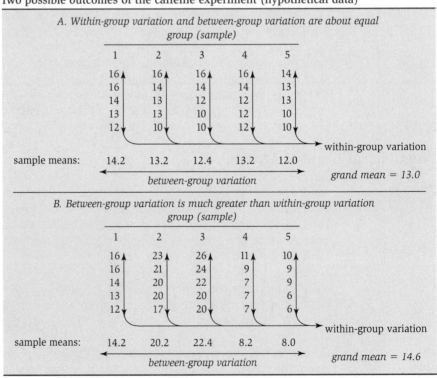

A. *Within-group variation and between-group variation are about equal*
group (sample)

	1	2	3	4	5
	16	16	16	16	14
	16	14	14	14	13
	14	13	12	12	13
	13	13	10	12	10
	12	10	10	12	10
sample means:	14.2	13.2	12.4	13.2	12.0

within-group variation

between-group variation

grand mean = 13.0

B. *Between-group variation is much greater than within-group variation*
group (sample)

	1	2	3	4	5
	16	23	26	11	10
	16	21	24	9	9
	14	20	22	7	9
	13	20	20	7	6
	12	17	20	7	6
sample means:	14.2	20.2	22.4	8.2	8.0

within-group variation

between-group variation

grand mean = 14.6

that the between-group variation and within-group variation are about equal, as would be expected if H_0 is true.

What if caffeine does have an effect? Suppose that one particular dosage causes an increase (or a decrease!) in test scores. This implies that it has about the same effect on all subjects receiving that dosage. The variability *within* that group will not be affected, since adding a constant to all scores or subtracting a constant from all scores does *not* change the variance or standard deviation. Thus the within-group variance estimate also will not be affected. The mean of this group, however, will be higher than the mean of the other groups. Therefore, the *between*-group variance estimate will increase. Similarly, if several different dosages have different effects on test scores (for example, some dosages increase test scores more than others), the within-group variance estimate will not increase but the between-group variance estimate will be larger. The greater the differences among the group means, the larger will be the between-group variance estimate.

As an illustration, consider the data in Table 15.1B. These data were obtained from the data in Table 15.1A as follows: scores in Group 1 were not changed; 7 was added to each score in Group 2; 10 was added to each score in Group 3; 5 was subtracted from each score in Group 4; and 4 was subtracted from each score in Group 5. The variance of each of the groups in Table 15.1B, and hence the within-group variance estimate, is equal to that of Table 15.1A. However, the between-group variance estimate is much greater in the case of Table 15.1B. Thus it is more likely that the five samples in Table 15.1B do *not* come from the same population.

To clarify further the meaning of between-group variation and within-group variation, consider the score of the first person in Group 1 of Table 15.1A, which is equal to 16. This person's score deviates by 1.8 points from the mean of Group 1. This is "error" in the sense that it cannot be explained by the caffeine variable, since all people in Group 1 receive the same dosage. It simply reflects the basic variance of the English test scores. The difference between the mean of Group 1 and the grand mean (14.2 − 13.0) is equal to 1.2 points. This reflects how much the dosage given Group 1 caused the mean test score of this group to differ from the grand mean, *plus* this same basic variance of the English test scores. Finally, the within-group variation of 1.8 and the between-group variation of 1.2 sum to 3.0, which equals the total variation of this score from the grand mean (16 − 13.0). The three deviation scores involving this person are:

1. total deviation = 3.0

2. within-group (from own group) deviation *(error)* = 1.8

3. between-group (from own group to grand mean) deviation = 1.2

The total variance, the within-group variance estimate, and the between-group variance estimate express the magnitude of these deviations, respectively, for everyone in the experiment. It can readily be shown that the total sum of squared deviations from the grand mean (the total sum of squares) is equal to the sum of the between-group sum of squares and the within-group sum of squares.*

The between-group variance estimate includes *both* the effects of the caffeine (if any) *and* error variance. The group means will be affected both by the experimental treatment, and by the basic English test or error variance of the scores on which they are based. The within-group variance estimate, however, reflects solely the error variance. The effect of the caffeine can therefore be determined by computing the following *F* ratio:†

$$F = \frac{\text{treatment variance} + \text{error variance}}{\text{error variance}}$$

$$= \frac{\text{between-group variance estimate}}{\text{within-group variance estimate}}$$

If H_0 is true, there is no treatment variance. The between-group and within-group variance estimates will therefore be approximately equal, so that *F* will be approximately equal to 1.0. The more *F* is greater than 1.0, the more sure you can be that caffeine does have an effect on test scores. You reject H_0 when *F* is so large as to have a probability of .05 or less of occurring if H_0 is true.

Values of *F* less than 1.0 would indicate that H_0 should be retained, since the between-group variance estimate is smaller than the within-group variance estimate. Since the value of *F* expected if H_0 is true is 1.0, one might conceive of a *significantly small* value of *F* (for example, 0.2). There would be no obvious explanation for such a result other than chance, or perhaps failure of one of the assumptions underlying the *F* test. Thus, you would retain H_0 in such instances, and check the possibility that some systematic factor like nonrandom sampling has crept in.

Computational Procedures

Sums of Squares

The first step in the ANOVA design is to compute the *sum of squares between groups* (symbolized by SS_B), the *sum of squares within groups* (symbolized by SS_W), and the *total sum of squares* (symbolized by SS_T). A *sum of squares* is nothing more than a sum of squared deviations, the numerator of the variance formula defined in Chapter 5.

1. Total sum of squares (SS_T).   The definition formula for the total sum of squares is

$$SS_T = \sum (X - \overline{X})^2$$

---

* The proof is given in the Appendix at the end of this chapter.

† If there are only two groups, the *F* test and the procedures given in Chapter 11 yield similar results. In fact, in this instance, $F = t^2$.

where

$$\overline{X} = \text{grand mean for all observations in the experiment}$$
$$\sum = \text{summation across all } observations$$

In Table 15.1A, for example,

$$SS_T = (16 - 13)^2 + (16 - 13)^2 + (14 - 13)^2$$
$$+ \cdots + (13 - 13)^2$$
$$+ (10 - 13)^2 + (10 - 13)^2$$
$$= 3^2 + 3^2 + 1^2 + \cdots + 0^2 + (-3)^2 + (-3)^2$$
$$= 100$$

As was the case with the variance (Chapter 5), however, it is usually easier to compute SS_T by using a shortcut computing formula:

$$SS_T = \sum X^2 - \frac{(\sum X)^2}{N}$$

where

$$N = \text{total number of observations in the experiment}$$

For the data in Table 15.1A,

$$\sum X^2 = 16^2 + 16^2 + 14^2 + \cdots 13^2 + 10^2 + 10^2$$
$$= 4{,}325$$
$$\frac{(\sum X)^2}{N} = \frac{(325)^2}{25} = 4{,}225$$
$$SS_T = 4{,}325 - 4{,}225$$
$$= 100$$

2. Sum of squares between groups (SS_B).   The definition formula for the sum of squares between groups is

$$SS_B = \sum N_G (\overline{X}_G - \overline{X})^2$$

where

$$N_G = \text{number of scores in group G}$$
$$\overline{X}_G = \text{mean of group G}$$
$$\overline{X} = \text{grand mean}$$
$$\sum = \text{summation across all the } groups \text{ (}not\text{ the subjects)}$$

As we have seen, the between-groups sum of squares deals with the difference between the mean of each group and the grand mean. The squared difference is effectively counted once for each person in the group. (Samples need not be of equal size.) In Table 15.1A,

$$SS_B = 5(14.2 - 13.0)^2 + 5(13.2 - 13.0)^2 + 5(12.4 - 13.0)^2$$
$$+ 5(13.2 - 13.0)^2 + 5(12.0 - 13.0)^2$$
$$= 5(1.2)^2 + 5(.2)^2 + 5(-.6)^2 + 5(.2)^2 + 5(-1.0)^2$$
$$= 14.4$$

A simpler computing formula for SS_B is

$$SS_B = \frac{(\sum X_1)^2}{N_1} + \frac{(\sum X_2)^2}{N_2} + \cdots + \frac{(\sum X_k)^2}{N_k} - \frac{(\sum X)^2}{N}$$

where

$\sum X_1$ = sum of scores in group 1, $\sum X_2$ = sum of scores in group 2, etc.
k = the number of groups (hence, the last group)
N_1 = number of scores in group 1, N_2 = number of scores in group 2, etc.

For example, in Table 15.1A,

$$SS_B = \frac{(71)^2}{5} + \frac{(66)^2}{5} + \frac{(62)^2}{5} + \frac{(66)^2}{5} + \frac{(60)^2}{5} - \frac{(325)^2}{25}$$
$$= 4,239.4 - 4,225$$
$$= 14.4$$

3. Sum of squares within groups (SS_W).   The definition formula for the sum of squares within groups is

$$SS_W = \sum (X_1 - \overline{X}_1)^2 + \sum (X_2 - \overline{X}_2)^2 + \cdots + \sum (X_k - \overline{X}_k)^2$$

where

X_1 = score in group 1, X_2 = score in group 2, etc.
k = last group
$\overline{X}_1$ = mean of first group, $\overline{X}_2$ = mean of second group, etc.
$\sum$ = summation across the N_G cases of the group in question

In Table 15.1A,

$$SS_W = (16 - 14.2)^2 + (16 - 14.2)^2 + (14 - 14.2)^2$$
$$+ (13 - 14.2)^2 + (12 - 14.2)^2 + (16 - 13.2)^2$$
$$+ (14 - 13.2)^2 + (13 - 13.2)^2 + \cdots$$
$$+ (10 - 12.0)^2 + (10 - 12.0)^2$$
$$= 85.6$$

That is, the difference between each score and the mean of the group containing the score is squared. Once this has been done for every score, the results are summed. This is the most tedious of the three sums of squares to compute, and it is possible to make use of the fact that

$$SS_T = SS_B + SS_W$$

That is, the sum of squares between groups and the sum of squares within groups must add up to the total sum of squares. (See the Appendix at the end of this chapter.) Therefore, the within-group sum of squares can readily be computed as follows:

$$SS_W = SS_T - SS_B$$
$$= 100 - 14.4$$
$$= 85.6$$

You should be careful in using this shortcut, however, because it provides no check on computational errors. If SS_W is found directly, the above formula can be used as a check on the accuracy in computation.

Mean Squares

Next, SS_B and SS_W are each divided by the appropriate degrees of freedom. The values thus obtained are called *mean squares,* and are estimates of the population variance. The degrees of freedom between groups (symbolized by df_B) are equal to

$$df_B = k - 1$$

where

$$k = \text{number of groups}$$

The degrees of freedom within groups (symbolized by df_w) are equal to

$$df_w = N - k$$

where

$$N = \text{total number of observations}$$

This is equivalent to obtaining the degrees of freedom for each group separately $(N_G - 1)$ and then adding the *df* across all groups. The *total* degrees of freedom, helpful as a check on the calculations of *df*, is equal to $N - 1$.

In Table 15.1A, $df_B = (5 - 1) = 4$; $df_W = (25 - 5) = 20$. The total $df = N - 1$ or 24, which is in fact equal to 4 + 20.

The mean squares between groups (symbolized by MS_B) and the mean squares within groups (symbolized by MS_W) are equal to:

$$MS_B = \frac{SS_B}{df_B}$$

$$MS_W = \frac{SS_W}{df_W}$$

Thus, in Table 15.1A,

$$MS_B = \frac{14.4}{4}$$

$$= 3.60$$

$$MS_W = \frac{85.6}{20}$$

$$= 4.28$$

The *F* Ratio

Having computed the mean squares, the last step is to compute the *F* ratio, where

$$F = \frac{MS_B}{MS_W}$$

In Table 15.1A,

$$F = \frac{3.60}{4.28}$$

$$= 0.84$$

Testing the *F* Ratio for Statistical Significance

As noted above, we encounter a new statistical model when we wish to test the ANOVA H_0 for statistical significance: the *F distributions*. Here again, the

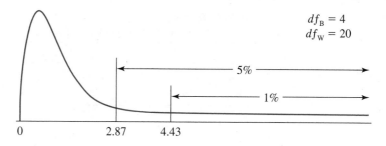

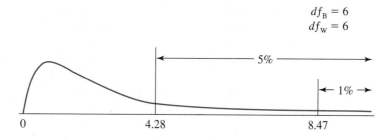

FIGURE 15.1

F distributions for $df_\mathrm{B} = 4$ and $df_\mathrm{W} = 20$, and $df_\mathrm{B} = 6$ and $df_\mathrm{W} = 6$

correct model enables us to determine the probability of obtaining the results observed in the samples *if* H_0 is true.

Just as there are different t distributions for different degrees of
freedom, so are there different F distributions for all combinations of different df_B and df_W. While the exact shape of an F distribution depends on
df_B and df_W, all F distributions are positively skewed. Two are illustrated in
Figure 15.1.

To obtain the minimum value of F needed to reject H_0, you refer to
Table F in the Appendix. Consult the *column* corresponding to df_B (the
degrees of freedom in the *numerator* of the F ratio), and the *row* corresponding to df_W (the degrees of freedom in the *denominator* of the F ratio).
The critical F values for 4 df (numerator) and 20 df (denominator) are:

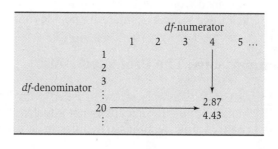

TABLE 15.2

Summary of one-way ANOVA of caffeine experiment

Source of Variation	SS	df	MS	F
Between groups	14.4	4	3.60	0.84
Within groups (error)	85.6	20	4.28	

The critical values for 20 df (numerator) and 4 df (denominator) are *not* the same; be sure to enter the table in the proper place. The smaller of the two values is the .05 criterion; the larger is the .01 criterion. In the case of Table 15.1A, the obtained value of 0.84 is less than the critical value of 2.87. Therefore, using $\alpha = .05$, you retain H_0 and conclude there is not sufficient reason to believe that there are any differences among the five population means. (Since the computed value of F is less than 1.0, you could have reached this conclusion without bothering to consult the table in this instance. The *expected* or mean value of any F distribution is 1, so no value of 1 or less can be significant.)

The ANOVA Table

It is customary to summarize the results of an analysis of variance in a table such as the one shown in Table 15.2. Note that the between-groups source of variation is listed first, and that the value of F is entered at the extreme right of the between-groups row.

One-Way ANOVA With Unequal Sample Sizes

An example of one-way ANOVA for samples of unequal size is shown in Table 15.3. The problem is whether or not students in three different eastern colleges in the United States differ in their attitudes toward student participation in determining college curricula. Three samples are randomly obtained from the three populations, and each person is given an attitude measure (the higher the score, the more positive the attitude). The statistical analysis is shown in Table 15.3; the computed value of F is 5.08. The critical value of F obtained from the F table for 2 and 24 degrees of freedom and $\alpha = .05$ is 3.40. Since the obtained F value is larger than the critical value obtained from the table, you reject H_0 and conclude that it is unlikely that the three samples come from populations with equal means.

Multiple Comparisons: The Protected t Test

An investigator is rarely satisfied with knowing that a set of k (three or more) population means is not equal, yet this is all that the rejection of the ANOVA null hypothesis signifies. It is usually necessary to make more spe-

TABLE 15.3

ANOVA analysis of attitudes of students from three colleges in the eastern United States to student participation in determining college curricula

Group 1	Group 2	Group 3	
15	17	6	$H_0: \mu_1 = \mu_2 = \mu_3$
18	22	9	$H_1: H_0$ is untrue
12	5	12	$\alpha = .05$
12	15	11	
9	12	11	
10	20	8	
12	14	13	
20	15	14	
	20	7	
	21		

$N_1 = 8 \quad N_2 = 10 \quad N_3 = 9 \quad N = 27$

$\sum X_1 = 108 \quad \sum X_2 = 161 \quad \sum X_3 = 91 \quad \sum X = 360 \quad \sum X^2 = 5{,}372$

$$SS_T = 5{,}372 - \frac{(360)^2}{27}$$

$$= 5{,}372 - 4{,}800 = 572$$

$$SS_B = \frac{(108)^2}{8} + \frac{(161)^2}{10} + \frac{(91)^2}{9} - \frac{(360)^2}{27}$$

$$= 4{,}970.21 - 4{,}800 = 170.21$$

$$SS_W = 572 - 170.21 = 401.79$$

Source of Variation	SS	df	MS	F
Between groups	170.21	$3 - 1 = 2$	$170.21/2 = 85.10$	$85.10/16.74 = 5.08$
Within groups (error)	401.79	$27 - 3 = 24$	$401.79/24 = 16.74$	

cific comparisons among the k means. Depending upon the structure and purpose of the experiment, these multiple comparisons may take many forms and are subject to many considerations.

Sometimes the investigator wishes to make all $k(k - 1)/2$ of the possible mean comparisons. At other times, only some of them may be needed (for example, when several experimental groups are each to be compared with a single control group). Another important issue is whether a given comparison was planned in advance of the research *(a priori)*, or whether it was suggested by the results *(post hoc,* also known as "data-snooping"). And we must be concerned with keeping the *experimentwise* error rate at a reasonably low level, while maintaining adequate power.

The area of multiple comparisons is therefore an exceedingly complex one, and even a modest survey of the popular methods is well beyond the scope of an elementary textbook. The interested reader is referred to the relatively simple treatments of some selected methods given by Kirk.[*]

Fortunately, there is available a simple method which is appropriate for most of the multiple comparison problems encountered in practice. This method has been demonstrated to perform remarkably well in an extensive comparative investigation over a wide variety of circumstances.[†] It is known as the Fisher LSD or "protected" t test procedure.

If *and only if* an ANOVA F test has resulted in the rejection of the overall null hypothesis that all k means are equal, any (or all) of the paired means may be compared by t tests using the usual decision rule. These t tests take advantage of the more stable estimate of the population variance provided by the ANOVA's MS_W, which is based on $df = N - k$. (Recall that an ordinary t test between $\overline{X}_i$ and $\overline{X}_j$ would be based on only the N_i and N_j observations, and would only have $df = N_i + N_j - 2$.) By requiring that the ANOVA F be significant, we protect the resulting ts from the large experimentwise Type I error rate that would otherwise occur. At the same time, these t tests are fairly powerful in detecting real population mean differences.

In the caffeine experiment, F was found to be nonsignificant, so no further analysis by t's is warranted. We will therefore illustrate the protected t procedure using the example of the ANOVA analysis of student attitudes given in Table 15.3. Since the F test for the ANOVA was significant at $\alpha = .05$, we may proceed to compare the three pairs of means by t tests. The null hypothesis for any pair of means, $\overline{X}_i$ and $\overline{X}_j$, is tested by

$$t = \frac{\overline{X}_i - \overline{X}_j}{\sqrt{MS_W \left(\frac{1}{N_i} + \frac{1}{N_j} \right)}}$$

Here, $df = N - k$, and MS_W is taken from the ANOVA results.

Table 15.4 gives the means, group sizes, and MS_W of Table 15.3 and illustrates the computation of these t's. For $df = 24$, the critical value for t at $\alpha = .05$ (two-tailed) is 2.064. Therefore, only the difference between the means of Groups 2 and 3 is significant. We can thus specify the rejection of the overall ANOVA null hypothesis (all three population means are not equal) by rejecting one of its composite pairwise null hypotheses, and asserting that $\mu_2 \neq \mu_3$.

When all groups are of the same size, $N_i = N_j$ and the denominator of the above t test becomes a constant for all the pairwise comparisons. It then becomes possible to determine what least significant difference (LSD) between means is necessary for significance, using any given decision rule:

---

[*] R. E. Kirk, *Experimental design: Procedures for the behavioral sciences,* 2nd ed. (Monterrey, Calif.: Brooks Cole, 1982), Chapter 3.

[†] S. G. Carmer and M. R. Swanson, "An evaluation of ten multiple comparison procedures by Monte Carlo methods," *Journal of the American Statistical Association* (1973): **68,** 66–74.

TABLE 15.4

Protected t tests among the three mean attitude scores following a significant ANOVA F (Table 15.3)

Means	$\overline{X}_1 = 13.50$ $N_1 = 8$	$\overline{X}_2 = 16.10$ $N_2 = 10$	$\overline{X}_3 = 10.11$ $N_3 = 9$

$$MS_W = 16.74$$

$\overline{X}_1$ versus $\overline{X}_2$: $\quad t = \dfrac{13.50 - 16.10}{\sqrt{16.74\left(\dfrac{1}{8} + \dfrac{1}{10}\right)}} = \dfrac{-2.60}{1.94} = -1.34$

$\overline{X}_1$ versus $\overline{X}_3$: $\quad t = \dfrac{13.50 - 10.11}{\sqrt{16.74\left(\dfrac{1}{8} + \dfrac{1}{9}\right)}} = \dfrac{3.39}{1.99} = 1.70$

$\overline{X}_2$ versus $\overline{X}_3$: $\quad t = \dfrac{16.10 - 10.11}{\sqrt{16.74\left(\dfrac{1}{10} + \dfrac{1}{9}\right)}} = \dfrac{5.99}{1.88} = 3.19$

df for t's $= df_W = N - k = 24$

$$\text{LSD} = t \sqrt{MS_W \left(\frac{2}{N_i}\right)}$$

The value of t is chosen for $df = N - k$ and the two-tailed α of the decision rule. Assume that all of the sample sizes were 10 ($= N_i$) in the preceding example, and the usual $\alpha = .05$ decision rule is to be used. Then MS_W ($= 16.74$) would be based on $df = N - k = 30 - 3 = 27$, and t for $\alpha = .05$ and $df = 27$ is found in Table C to equal 2.052. Then,

$$\text{LSD} = 2.052 \sqrt{16.74 \left(\frac{2}{10}\right)} = 2.052(1.83) = 3.75$$

All pairs of means differing by at least 3.75 attitude scale points would be significantly different at $\alpha = .05$ (two-tailed).

Although the protected t procedure provides good control and balance of Type I and Type II errors for most applications in the behavioral sciences, for certain patterns of population means the protection afforded by requiring F to be significant wanes as k increases. Some other procedure should be used when k is large (say, six or more), or when exact control of the experimentwise Type I error rate is desired. Also, it is possible (though unlikely) to find F significant and none of the pairwise ts significant, in which case the proper conclusion is that the k means are not all equal, but the data do not justify any further specification (*not* that all paired means are equal, which would be a contradiction).

Confidence Intervals for the Protected t Test

In the previous section, we tested the null hypothesis for each pair of means using three protected t tests. An alternative procedure would be to calculate three confidence intervals, as discussed in Chapters 10 and 11. This has the advantage of specifying all reasonably likely values of the difference between the two population means in each comparison.

To use the confidence interval procedure for the protected t test, the ANOVA F test must have resulted in the rejection of the overall null hypothesis that all k means are equal. As we observed in Chapter 11, the confidence interval for the difference between two population means is:

$$[(\overline{X}_1 - \overline{X}_2) - ts_{\overline{X}_1 - \overline{X}_2}] \leq \mu_1 - \mu_2 \leq [(\overline{X}_1 - \overline{X}_2) + ts_{\overline{X}_1 - \overline{X}_2}]$$

To calculate the critical value of t, you must use the correct degrees of freedom. For the data illustrated in Table 15.4, $df = N - k$ or 24, the degrees of freedom associated with the within-groups mean square (MS_W). For $df = 24$ and $\alpha = .05$, and a two-tailed test, the critical value of t is 2.052.

For each comparison, the standard error of the difference is equal to $\sqrt{MS_W(1/N_i + 1/N_j)}$. Since $MS_W = 16.74$, $N_1 = 8$, $N_2 = 10$, and $N_3 = 9$, the three standard errors of the difference that we need are:

(1) Group 1 versus Group 2: $\sqrt{16.74(1/8 + 1/10)} = 1.94$

(2) Group 1 versus Group 3: $\sqrt{16.74(1/8 + 1/9)} = 1.99$

(3) Group 2 versus Group 3: $\sqrt{16.74(1/10 + 1/9)} = 1.88$

Therefore, the three confidence intervals are:

(1) Group 1 versus Group 2:

$$-2.60 - (2.052)(1.94) \leq \mu_1 - \mu_2 \leq -2.60 + (2.052)(1.94)$$
$$-6.58 \leq \mu_1 - \mu_2 \leq 1.38$$

(2) Group 1 versus Group 3:

$$3.39 - (2.052)(1.99) \leq \mu_1 - \mu_3 \leq 3.39 + (2.052)(1.99)$$
$$-0.69 \leq \mu_1 - \mu_3 \leq 7.47$$

(3) Group 2 versus Group 3:

$$5.99 - (2.052)(1.88) \leq \mu_2 - \mu_3 \leq 5.99 + (2.052)(1.88)$$
$$2.13 \leq \mu_2 - \mu_3 \leq 9.85$$

As would be expected from the results in Table 15.4, only for the comparison between groups 2 and 3 does zero fall outside the confidence interval. Here again, confidence intervals provide us with important additional information about the probable difference between the population means in each comparison.

Some Comments on the Use of ANOVA

Underlying Assumptions

The procedures discussed in this chapter assume that observations are *independent*. The value of any observation should *not* be related in any way to that of any other observation (as would be the case if the three samples were selected to be equal on average IQ). It is also formally assumed that the variances are equal for all treatment populations (that is, homogeneous), and that the populations are normally distributed. ANOVA is robust with regard to these assumptions, however, and will yield accurate results even if population variances are not homogeneous (provided that sample sizes are about equal) and even if population shapes depart moderately from normality.

A Measure of Strength of Relationship

As was pointed out in previous chapters, statistical significance does not necessarily imply a strong relationship. It is therefore desirable to have an overall measure of the strength of the relationship between the independent and dependent variable, in addition to the test of significance. One such measure is provided by ε (epsilon):

$$\varepsilon = \sqrt{\frac{df_B\,(F-1)}{df_B\,F + df_W}}$$

Consider the data in Table 15.3, where $F = 5.08$, $df_B = 2$, and $df_W = 24$. Then, ε is equal to:

$$\varepsilon = \sqrt{\frac{(2)(5.08 - 1)}{(2)\,5.08 + 24}}$$

$$= \sqrt{\frac{8.16}{34.16}}$$

$$= \sqrt{.23887}$$

$$= .49$$

Epsilon bears the same relationship to F that r_{pb} bears to t (see Chapter 13). Thus, you would conclude that there is a fairly strong relationship in the sample between college and attitudes to student participation in determining college curricula. Its significance is, of course, given by the F.

Summary

Analysis of variance (ANOVA) permits null hypotheses to be tested which involve the means of three or more samples (groups). One-way ANOVA deals with one independent variable made up of membership in one of k groups or levels. Total variance is partitioned into two sources: *between-group variance* and *within-group (error) variance.* These are compared by using the F ratio to determine whether or not the independent variable has an effect on the dependent variable. A significant F ratio in ANOVA indicates that not all k means are equal. A follow-up procedure (protected t test) is used to specify which pairs of means differ significantly.

1. Meaning of symbols

Symbol	General Meaning
k	Number of groups (or the last group)
N	Total number of observations
N_G	Number of observations in group G
$\overline{X}$	Grand mean
$\overline{X}_G$	Mean of group G
X_G	A score in group G

2. Definition and computing formulas

1. Total sum of squares

    Definition: $SS_T = \sum (X - \overline{X})^2$

    Computing formula: $SS_T = \sum X^2 - \dfrac{(\sum X)^2}{N}$

2. Between-groups sum of squares

    Definition: $SS_B = \sum N_G(\overline{X}_G - \overline{X})^2$

    Computing formula: $SS_B = \dfrac{(\sum X_1)^2}{N_1} + \dfrac{(\sum X_2)^2}{N_2} + \cdots$

    $\qquad\qquad + \dfrac{(\sum X_k)^2}{N_k} - \dfrac{(\sum X)^2}{N}$

3. Within-groups sum of squares

    Definition: $SS_W = \sum (X_1 - \overline{X}_1)^2 + \sum (X_2 - \overline{X}_2)^2$

    $\qquad\qquad + \cdots + \sum (X_k - \overline{X}_k)^2$

    Computing formula: $SS_W = SS_T - SS_B$

3. Steps in one-way analysis of variance

1. Compute SS_T, SS_B, and SS_W.

2. Compute
$$df_B = k - 1$$
$$df_W = N - k$$

3. Compute mean squares (MS):
$$MS_B = \frac{SS_B}{df_B}$$
$$MS_W = \frac{SS_W}{df_W}$$

4. Compute
$$F = \frac{MS_B}{MS_W}$$
$$df = k - 1, N - k$$

5. Obtain the critical value from the F table and test for significance. If the computed F is equal to or greater than the tabled F, reject H_0 (that all population means are equal) in favor of H_1 (that H_0 is not true). Otherwise, retain H_0.

6. *If F is statistically significant,* multiple comparisons may be run to determine which of the population means differ from one another. Though this is a complex area, one technique that will usually be appropriate is the Fisher *LSD* (or "protected" *t* test) procedure. For any (or every) pair of groups, compute:
$$t = \frac{\overline{X}_i - \overline{X}_j}{\sqrt{MS_W \left(\frac{1}{N_i} + \frac{1}{N_j} \right)}}, \qquad df = N - k$$

where
$$\overline{X}_i = \text{mean of group } i$$
$$\overline{X}_j = \text{mean of group } j$$
$$MS_W = \text{within-groups mean square}$$
$$N_i = \text{number of observations in group } i$$
$$N_j = \text{number of observations in group } j$$

Do *not* use this procedure if k, the number of *groups*, is greater than about 6–8.

Confidence intervals may also be determined for each of the comparisons in the protected t test. Such intervals provide more information than does the null-hypothesis test, since they specify all reasonably likely values of the difference between the two population means in each comparison.

4.   *Measure of strength of relationship*

One measure for determining the strength of the relationship between the independent and dependent variables is ε (epsilon):

$$\varepsilon = \sqrt{\frac{df_B \, (F - 1)}{df_B \, F + df_W}}$$

Epsilon bears the same relationship to F that r_{pb} bears to t (see Chapter 13).

Appendix: Proof That the Total Sum of Squares Is Equal to the Sum of the Between-Group and the Within-Group Sum of Squares

It is obvious that

$$X - \overline{X} = X - \overline{X} + \overline{X}_G - \overline{X}_G$$

Rearranging the terms,

$$X - \overline{X} = (X - \overline{X}_G) + (\overline{X}_G - \overline{X})$$

Squaring both sides of the equation and summing over *all* people gives

$$\sum (X - \overline{X})^2 = \sum (X - \overline{X}_G)^2 + \sum (\overline{X}_G - \overline{X})^2$$
$$+ \, 2 \sum (X - \overline{X}_G)(\overline{X}_G - \overline{X})$$

For any one group, $(\overline{X}_G - \overline{X})$ is a constant, and the sum of deviations about the group mean must equal zero. The last term is therefore always equal to zero, leaving

$$\sum (X - \overline{X})^2 = \sum (X - \overline{X}_G)^2 + \sum (\overline{X}_G - \overline{X})^2$$

For ANOVA purposes, the sums of squares are then divided by the appropriate degrees of freedom to yield estimates of the population variance, which are then compared by the F ratio.

Chapter 16

Introduction to Factorial Design: Two-Way Analysis of Variance

PREVIEW

Introduction

When should you use a factorial design instead of a one-way analysis of variance? In what way are these two procedures similar? What are some of the important differences between them?

What is meant by the interaction between two variables?

Computational Procedures

What are the procedures for computing a two-way analysis of variance (two-way factorial design)?

What is the correct statistical model to use in this situation?

How do we test the F ratios for statistical significance?

What are the multiple comparison procedures in two-way analysis of variance?

The Meaning of Interaction

What is meant by a zero interaction and a large interaction between two variables?

What are two possible types of interaction for a 2×2 factorial design?

How do we interpret the results when the interaction is significant and neither, or either, or both of the factors are significant?

Summary

Introduction

The one-way analysis of variance presented in Chapter 15 is used to investigate the relationship of a *single* independent variable to a dependent variable, where the independent variable has two or more levels (that is, groups). For example, the caffeine experiment in Chapter 15 dealt with the effect of five different dosages of caffeine (five levels of the independent variable) on performance on an English test (the dependent variable).

The *factorial design* is used to study the relationship of *two or more* independent variables (called *factors*) to a dependent variable, where each factor has two or more levels. Suppose you are interested in the relationship between four different dosages of caffeine (four levels: large, moderate, small, zero) and sex (two levels: male and female) to scores on a 20-item English test. There are several hypotheses of interest: different dosages of caffeine may affect test scores; males and females may differ in test performance; certain caffeine dosages may affect test scores more for one sex than for the other. These hypotheses may be evaluated in a single statistical framework by using a factorial design. This example would be called a "two-way" analysis of variance, since there are two independent variables. It could also be labeled as a 4×2 factorial design, because there are four levels of the first independent variable and two levels of the second independent variable.

The logic of the factorial design follows from the logic of the more simple one-way design. The total sum of squares is partitioned into within-group (error) sum of squares and between-group sum of squares. In the factorial design, however, the between-group sum of squares is itself partitioned into several parts: variation due to the first factor, variation due to the second factor, and variation due to the joint effects of the two factors (called the *interaction*). (See Figure 16.1.) An example of an interaction effect would be if a particular dosage of caffeine improved test scores for males but *not* for females, while other dosages had no effect on test scores for either sex. (The

FIGURE 16.1

Partitioning of variation in a two-way factorial design

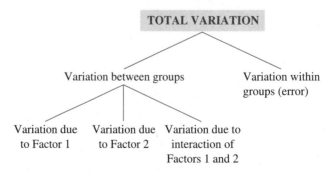

interaction, a particularly important feature of factorial design, will be discussed in detail later in this chapter.)

It can be shown algebraically that the total sum of squares is equal to the sum of these various parts: (1) the within-group (error) sum of squares, (2) the sum of squares due to factor 1, (3) the sum of squares due to factor 2, and (4) the sum of squares due to the interaction of factors 1 and 2. Thus a factorial design makes it possible to break down the total variability into *several* meaningful parts. That is, it permits several possible explanations as to why people are different on the dependent variable. As was pointed out in Chapter 5, explaining variation—why people differ from one another—is the *raison d'être* of the behavioral scientist.

Computational Procedures

The outline of the computational procedure for the two-way factorial design is as follows:

1. A. Compute SS_T.
   B. Compute SS_B.
   C. Subtract SS_B from SS_T to obtain SS_W (error).
   D. Compute SS_1 (the sum of squares for factor 1).
   E. Compute SS_2 (the sum of squares for factor 2).
   F. Subtract SS_1 and SS_2 from SS_B to obtain the sum of squares for the interaction of factors 1 and 2 ($SS_{1\times2}$).

2. Convert the sums of squares in C, D, E, and F above to mean squares by dividing each one by the appropriate number of degrees of freedom.

3. A. Test the mean square for factor 1 (MS_1) for statistical significance by computing the appropriate F ratio.
   B. Test the mean square for factor 2 (MS_2) for statistical significance by computing the appropriate F ratio.
   C. Test the mean square for interaction ($MS_{1\times2}$) for statistical significance by computing the appropriate F ratio.

The two-way factorial design permits you to test three null hypotheses—one concerning the effect of factor 1, one concerning the effect of factor 2, and one concerning the *joint effect* of factor 1 and factor 2—in a single statistical framework.

As an illustration, consider the hypothetical results for the caffeine experiment shown in Table 16.1. There are five observations in each cell; for example, the scores of 6, 15, 12, 12, and 13 are the test scores of five men who received a large dose of caffeine. Just as in the one-way design, the within-group variance estimate is based on the variability *within* each of the eight cells. Variation due to the caffeine factor is reflected by the variability across the four *column* means, while variation due to the sex factor is reflected by the variability (that is, difference) of the two *row* means.

TABLE 16.1

Scores on a 20-item English test as a function of caffeine dosage and sex (4×2 factorial design)

		Caffeine Dosage (Factor 1)									Row Sums	Row Means
		Large		Moderate		Small		Zero				
Male	6 15 12 12 13	(sum = 58)	12 10 12 13 7	(sum = 54)	10 13 15 12 10	(sum = 60)	9 10 7 12 7	(sum = 45)			217	10.85
Sex (Factor 2)												
Female	10 13 4 9 5	(sum = 41)	9 7 10 7 13	(sum = 46)	12 13 15 10 13	(sum = 63)	4 7 6 9 9	(sum = 35)			185	9.25
Column sums	99		100		123		80				Grand sum = 402	
Column means	9.9		10.0		12.3		8.0				Grand mean = 10.05	

Sums of Squares

1. Total sum of squares (SS_T). The total sum of squares is computed in the same way as in Chapter 15:

$$SS_T = \sum X^2 - \frac{(\sum X)^2}{N}$$

where

$$N = \text{total number of observations}$$
$$\sum = \text{summation across } observations$$

In Table 16.1,

$$\sum X^2 = 6^2 + 15^2 + 12^2 + \cdots + 6^2 + 9^2 + 9^2$$

$$= 4{,}394$$

$$\frac{(\sum X)^2}{N} = \frac{(402)^2}{40} = 4{,}040.1$$

$$SS_T = 4{,}394 - 4{,}040.1$$

$$= 353.9$$

2. Sum of squares between groups (SS_B). We can ignore for a moment the fact that this is a factorial design, treat the data in Table 16.1 as 8 groups, and find SS_B as in Chapter 15:

$$SS_B = \frac{(\sum X_1)^2}{N_1} + \frac{(\sum X_2)^2}{N_2} + \cdots + \frac{(\sum X_k)^2}{N_k} - \frac{(\sum X)^2}{N}$$

For the data in Table 16.1, $k = 8$, and

$$SS_B = \frac{(58)^2}{5} + \frac{(54)^2}{5} + \cdots + \frac{(63)^2}{5} + \frac{(35)^2}{5} - \frac{(402)^2}{40}$$

$$= \frac{20,896}{5} - \frac{161,604}{40}$$

$$= 4,179.2 - 4,040.1$$

$$= 139.1$$

3. Sum of squares within groups (error) (SS_w).   The within-groups sum of squares may be found by subtraction:

$$SS_W = SS_T - SS_B$$

$$= 353.9 - 139.1$$

$$= 214.8$$

4. Sum of squares for factor 1 (SS_1).   Let us define caffeine as factor 1. The sum of squares for the caffeine factor, *which ignores sex differences*, is*

$$SS_1 = \sum \frac{\textbf{(sum of each \textit{column})}^2}{\textbf{(\textit{N} in each column)}} - \frac{(\sum X)^2}{N}$$

where the first $\sum$ is summation across all *columns* (and $\sum X$ is over all N observations).

If caffeine has an effect on test scores (ignoring sex), the means (and hence the sums) of the *columns* of Table 16.1 should show high variability. The value of $(\sum X)^2/N$ has already been found to be equal to 4,040.1. Then,

$$SS_1 = \frac{(99)^2}{10} + \frac{(100)^2}{10} + \frac{(123)^2}{10} + \frac{(80)^2}{10} - 4,040.1$$

$$= 4,133 - 4,040.1$$

$$= 92.9$$

5. Sum of squares for factor 2 (SS_2).   Let us define sex as factor 2. The sum of squares for the sex factor, *which ignores differences in caffeine dosage*, is

$$SS_2 = \sum \frac{\textbf{(sum of each \textit{row})}^2}{\textbf{(\textit{N} in each row)}} - \frac{(\sum X)^2}{N}$$

---

* The formulas for many of the terms in factorial designs involve procedures of double and triple summation. Rather than introduce this notation solely for this chapter, however, we will use a more verbal presentation.

where the first $\sum$ is summation across all *rows* (and $\sum X$ is over all N observations). If sex has an effect on test scores (ignoring caffeine), the means (and hence the sums) of the *rows* of Table 16.1 should show high variability.

$$SS_2 = \frac{(217)^2}{20} + \frac{(185)^2}{20} - 4{,}040.1$$

$$= 4{,}065.7 - 4{,}040.1$$

$$= 25.6$$

6. Sum of squares for interaction ($SS_{1\times2}$).   The interaction sum of squares is part of the variability of the 8 cells and is obtained by subtraction:

$$SS_{1\times2} = SS_B - SS_1 - SS_2$$

In Table 16.1,

$$SS_{1\times2} = 139.1 - 92.9 - 25.6$$

$$= 20.6$$

Mean Squares

The next step is to convert each sum of squares to an estimate of the population variance, or mean square. This is done by dividing by the appropriate degrees of freedom, as shown below and illustrated in Figure 16.2.

Source	Degrees of Freedom	Computation for Table 16.1
Total	$N - 1$	$df_T = 40 - 1 = 39$
Within groups	$N - k$	$df_W = 40 - 8 = 32$
Between groups	$k = 1$	$df_B = 8 - 1 = 7$
Factor 1	(number of levels of factor 1) $- 1$	$df_1 = 4 - 1 = 3$
Factor 2	(number of levels of factor 2) $- 1$	$df_2 = 2 - 1 = 1$
interaction	$df_1 \times df_2$	$df_{1\times2} = 3 \times 1 = 3$

Note: N = total number of observations
      k = number of cells

The first three values in the preceding table are the same as in the case of the one-way design (Chapter 15). The total degrees of freedom equals $N - 1$, or one less than the total number of observations. The within-group degrees of freedom is equal to $N - k$, where k equals the number of cells (or groups). This is equivalent to obtaining the degrees of freedom for each cell (one less than the number of observations in the cell, or 4), and summing over all cells ($4 \times 8 = 32$). The between-group degrees of freedom equals one less than the number of cells.

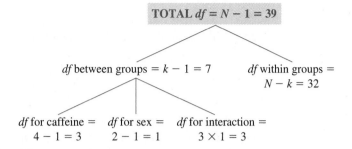

TOTAL $df = N - 1 = 39$

df between groups = $k - 1 = 7$ df within groups = $N - k = 32$

df for caffeine = df for sex = df for interaction =
$4 - 1 = 3$ $2 - 1 = 1$ $3 \times 1 = 3$

FIGURE 16.2

Partitioning of degrees of freedom in caffeine experiment

It is also necessary, however, to partition the between-group degrees of freedom in the same way as the between-group variance is partitioned. The degrees of freedom for factor 1 is one less than the number of levels of factor 1. In Table 16.1, there are four dosages of caffeine, so the degrees of freedom are $4 - 1 = 3$. Similarly, the degrees of freedom for factor 2, sex, are $2 - 1 = 1$. The degrees of freedom for the interaction of caffeine and sex is found by *multiplying* the degrees of freedom for each factor ($3 \times 1 = 3$).

The mean squares are then found by dividing each sum of squares by the corresponding degrees of freedom:

Mean square within groups:

$$(MS_W) = \frac{SS_W}{df_W} = \frac{214.8}{32} = 6.71$$

Mean square for caffeine:

$$(MS_1) = \frac{SS_1}{df_1} = \frac{92.9}{3} = 30.97$$

Mean square for sex:

$$(MS_2) = \frac{SS_2}{df_2} = \frac{25.6}{1} = 25.6$$

Mean square for interaction:

$$(MS_{1 \times 2}) = \frac{SS_{1 \times 2}}{df_{1 \times 2}} = \frac{20.6}{3} = 6.87$$

F Ratios and Tests of Significance

As with the one-way analysis of variance, the correct statistical model to use with factorial designs is the F distributions. The first null hypothesis to be tested is that the four caffeine groups come from populations with equal means. The mean square for the caffeine factor is divided by the mean square within groups, yielding the following F ratio:

$$F = \frac{MS_1}{MS_W} = \frac{30.97}{6.71} = 4.62$$

The critical value from the F table for three degrees of freedom in the numerator and 32 degrees of freedom in the denominator and $\alpha = .05$ is 2.90. Since the computed F of 4.62 is greater than this value, you reject H_0 and conclude that the four groups do *not* come from populations with equal means—that is, different dosages of caffeine *do* have an effect on the English test scores.

The second null hypothesis to be tested is that the two sexes come from populations with equal means. The mean square for the sex factor is divided by the mean square within groups, yielding the following F ratio:

$$F = \frac{MS_2}{MS_W} = \frac{25.60}{6.71} = 3.82$$

The critical value from the F table for one degree of freedom in the numerator and 32 degrees of freedom in the denominator and $\alpha = .05$ is 4.15. (Note that the critical values for the various F tests differ if the df differ, since they are different F distributions.) Since the computed value of 3.82 is less than the critical value, you retain H_0 and conclude that there is not sufficient reason to believe that different sexes perform differently on the English test.

The third null hypothesis to be tested is that the interaction effect is zero:

$$F = \frac{MS_{1 \times 2}}{MS_W} = \frac{6.87}{6.71} = 1.02$$

This computed value of F is less than the critical value obtained from the F table of 2.90 for 3 and 32 df and $\alpha = .05$. So you retain H_0, and conclude that there is not sufficient reason to reject the null hypothesis of no interaction effect.

ANOVA Table

The results of the above analysis of variance are summarized in Table 16.2. Note that the factors are identified by name for the convenience of the reader. Also, as was the case with the one-way analysis of variance, within-group variation (error) is listed last. No F value is listed for error because error is used as the denominator of the various F ratios, and is not itself the subject of a statistical test.

TABLE 16.2

Summary of two-way ANOVA of caffeine experiment

Source	SS	df	MS	F
Caffeine	92.90	3	30.97	4.62
Sex	25.60	1	25.60	3.82
Caffeine × sex	20.60	3	6.87	1.02
Error	214.80	32	6.71	

Protected *t* Tests in Factorial Design

We saw in one-way ANOVA how we could specify the meaning of a significant *F* ratio by means of the protected *t* procedure. In a two-way factorial design, there are potentially three *F* tests that may require further specification as to which means differ from which others, and the same procedure may be followed.

When the *F* for interaction is *not* significant, it is meaningful to follow up each of the independent variables (factors) as if it were from a one-way ANOVA. If the *F* for a factor is *not* significant, there is nothing to follow up. Doing so is not only unnecessary but improper, since the resulting *t*'s would not be protected. Or if a significant factor has only two levels, there is nothing to follow up. There is only one difference, so it must be significant. But a significant *F* for a factor of three or more levels requires specification just as in the one-way design.

In the above example, the interaction *F* is not significant and the *F* for caffeine is. So you may proceed to perform protected *t* tests among the four caffeine dosage level means on the English test. Table 16.1 gives these as 9.9, 10.0, 12.3, and 8.0. We found in the ANOVA that $MS_W = 6.71$, based on $df = 32$. Since all of the levels have the same $N_i = 10$, the critical *t* for $\alpha = .05$ at $df = 32$ (Table C) is approximately 2.04. Then, use the formula for the least significant difference given in Chapter 15:

$$\text{LSD} = t \sqrt{MS_W \left(\frac{2}{N_i} \right)}$$

$$= 2.04 \sqrt{6.71 \left(\frac{2}{10} \right)} = 2.36$$

Differences between dosage level means as large as 2.36 are significant at $\alpha = .05$. By this criterion, the mean for the small dosage condition (12.3) is significantly larger than the means for the large dosage (9.9) and zero dosage (8.0) conditions. No other mean difference is significant.

Alternatively, a confidence interval may be established for each comparison. The same restriction applies, namely that the overall *F* for the factor must be significant, and the procedure is the same as for the one-way analysis of variance.

In the caffeine experiment, the critical value of t for $\alpha = .05$ is 2.04. Since there are 10 cases in each group, the standard error of the difference is the same for all comparisons. It is equal to $\sqrt{6.71(1/10 + 1/10)}$ or 1.16. Thus, the 95% confidence interval for the difference between Group 1 (large dosage) and Group 3 (small dosage) is:

$$[(9.9 - 12.3) - (2.04)(1.16)] \leqslant \mu_1 - \mu_3 \leqslant [(9.9 - 12.3) + (2.04)(1.16)]$$

$$-4.76 \leqslant \mu_1 - \mu_3 \leqslant -0.04$$

The value of zero falls outside this interval, indicating that this comparison is statistically significant. In addition, the confidence interval specifies all reasonably likely values of the difference between the two population means. Other confidence intervals for the caffeine factor may be obtained in a similar fashion.

When the F for interaction *is* significant, it indicates that the factors operate *jointly*. (See the following section.) Under these circumstances, our interest is drawn to differences between *cells*, or specific combinations of the two factors. In the previous example, the F for interaction was not significant, so cell means would not be compared. But an alternative set of results for this experiment, illustrating a large interaction, is posited in Table 16.4 in the next section. For these results, the MS for caffeine $\times$ sex is 34.87. Assuming the same $MS_W = 6.71$ ($df = 32$) as before, the F for the caffeine $\times$ sex interaction is $34.87/6.71 = 5.19$. This is significant, since it exceeds the critical value of the F table for 3 and 32 df and $\alpha = .01$, which is 4.46.

With a significant interaction F ratio, we can now perform protected t's on differences in the *cell* means of Table 16.4B. Since all cells have the same $N_i = 5$, we can find the LSD. Using the same $MS_W = 6.71$ and (for $\alpha = .05$ at $df = 32$) the same critical $t = 2.04$, we find that:

$$\text{LSD} = 2.04 \sqrt{6.71 \left(\frac{2}{5} \right)} = 3.34$$

All differences between cell means as large as 3.34 are significant at $\alpha = .05$. Turning to Table 16.4B, we find some interesting results (for these fictitious

TABLE 16.3

Cell means for the data in Table 16.1

		Caffeine Dosage				
		Large	Moderate	Small	Zero	Overall
Sex	Male	11.6	10.8	12.0	9.0	10.85
	Female	8.2	9.2	12.6	7.0	9.25
$\overline{X}_{male} - \overline{X}_{female}$		3.4	1.6	-0.6	2.0	1.60

data). Males show significantly higher English test means than females under the large (11.6 − 8.2 = 3.4) and the small (12.0 − 7.0 = 5.0) caffeine dosage conditions, but a significantly *lower* mean under the placebo (zero dosage) condition (9.0 − 12.6 = − 3.6). Further, while there are *no* significant differences for males among the four dosage levels, females show significantly higher performance under the zero dosage condition (12.6) than they do under each of the other three conditions (8.2, 9.2, and 7.0). Formally, other cell means may be compared (for example, large dosage–male versus zero dosage–female), but we restrict ourselves to those that make sense.

The Meaning of Interaction

Interaction refers to the *joint* effect of two or more factors on the dependent variable. To illustrate, let us return to the caffeine experiment. The interaction effect of caffeine and sex refers to the effect of particular joint combinations of the two factors, such as male-high dosage or female-moderate dosage, and *not* to the sum of the separate effects of the two factors. It is the joint effect *over and above* the sum of the separate effects.

A numerical illustration of interaction is given in Table 16.3, which summarizes the cell means for the caffeine experiment. The overall mean difference between the sexes, ignoring the caffeine factor, shows that the men averaged 1.6 points higher than the women. If the mean differences between males and females for each dose of caffeine were all equal to 1.6, the

TABLE 16.4

Hypothetical cell means illustrating zero and large interaction effects

A. Zero interaction

		Caffeine Dosage				
		Large	Moderate	Small	Zero	Overall
Sex	Male	11.6	10.8	12.0	9.0	10.85
	Female	10.0	9.2	10.4	7.4	9.25
$\overline{X}_{male} - \overline{X}_{female}$		1.6	1.6	1.6	1.6	1.60

B. Large interaction

		Caffeine Dosage				
		Large	Moderate	Small	Zero	Overall
Sex	Male	11.6	10.8	12.0	9.0	10.85
	Female	8.2	9.2	7.0	12.6	9.25
$\overline{X}_{male} - \overline{X}_{female}$		3.4	1.6	5.0	− 3.6	1.60

interaction effect would be zero. That is, sex differences in English test performance would be exactly the same regardless of the dosage of caffeine. Also, the difference between mean test scores for any two dosages of caffeine would be the same for both sexes. (See Table 16.4A.) If the mean differences for each dose of caffeine were considerably different from one another, there would be a significant interaction effect; the caffeine factor would affect test scores differently for different sexes. (One possible illustration is shown in Table 16.4B.) Since the interaction effect for the data in Table 16.3 was found to be *not* significant, the differences among the observed mean differences of 3.4, 1.6, -0.6, and 2.0 are likely to have occurred by chance (random sampling error). So there is *not* sufficient reason to believe that there is an interaction effect in the population from which the samples in this experiment were drawn.

A graphic illustration of the examples in Table 16.4 is shown in Figure 16.3. When the interaction effect is zero (Figure 16.3A), the line connecting the points corresponding to the cell means for men follows the same pattern as the corresponding line for women. That is, the distance between the two lines is the same at all points. When there is an interaction effect (Figure 16.3B), sex differences are different from one point to another. Using the LSD test for this example in the preceding section, we saw that while males had significantly higher means under large and small dosage conditions, females had a significantly higher mean under the zero dosage condition.

Two different kinds of interaction are shown for a 2 × 2 factorial design in Figure 16.4. In Figure 16.4A, men obtain higher average test scores than women when caffeine is administered, while women obtain higher average scores than men when the placebo is given. In Figure 16.4B, caffeine has little if any effect for women (the cell means for the caffeine and placebo conditions are about equal), but men perform better on the test when given caffeine. That is, sex differences are greater under the caffeine condition than

FIGURE 16.3

Graphic representation of data in Table 16.4

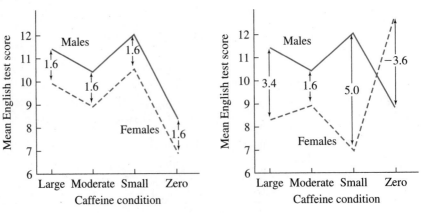

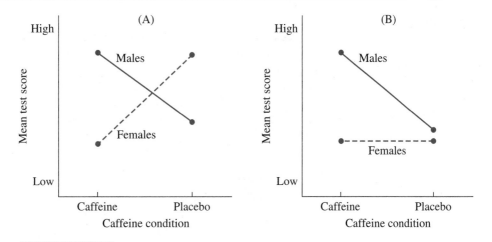

Two kinds of interaction effect for caffeine and sex on test scores (2 × 2 factorial design)

under the placebo condition. Thus, there is a "reversal" effect in Figure 16.4A; whereas in Figure 16.4B, caffeine affects test scores only for one sex. In practice, such assertions would need to be supported by protected t tests (or by some other multiple comparison procedure).

Just as the effects of sex and dosage level are independent of each other in this design, so is the interaction independent of each of the factors. This mutual independence means that any combination of these three kinds of effects can prove to be significant. For example, the interaction can prove to be significant when neither, or either, or both of the factors are significant. Thus, the *average* effect of dosage level (combining sexes) and the *average* effect of sex (combining dosage levels) might be zero, while males might be much higher than females at two dosage levels and much lower at two others. This would result in neither factor significant, but the interaction significant. Or, dosage level might result in different mean scores (combining sexes), while sexes (combining dosage levels) are the same *on the average.* The sex *difference* may still markedly favor males at one dosage level and slightly favor females at the other three, resulting in a significant interaction.

Much more could be said about interaction, and about factorial design. There are many other forms of factorial design: procedures exist for including three or more independent variables in a single design, or for including several scores obtained from the same people. A discussion of such techniques, however, is best reserved for advanced tests. Suffice it to say that the factorial design in its various forms, including as it does the important concept of interaction, is one of the most valuable statistical tools at the disposal of the behavioral scientist.

Summary

In two-way ANOVA (or two-way *factorial design*), one of many complex forms of ANOVA, there are *two* independent variables. The statistical analysis makes possible a significance test (using the F ratio) of the effect of *each* independent variable, and of the effect of the *interaction* of the two variables — that is, the *joint* effect of the two variables over and above the separate effects of each one.

1. Sums of squares

1. Total sum of squares (SS_T)

    Computing formula: $SS_T = \sum X^2 - \dfrac{(\sum X)^2}{N}$

2. Between-groups sum of squares (SS_B)

    Computing formula:

    $$SS_B = \frac{(\sum X_1)^2}{N_1} + \frac{(\sum X_2)^2}{N_2} + \cdots$$
    $$+ \frac{(\sum X_k)^2}{N_k} - \frac{(\sum X)^2}{N}$$

3. Within-groups sum of squares (SS_W)

    Computing formula: $SS_W = SS_T - SS_B$

4. Sum of squares for factor 1 (SS_1):

    Computing formula:

    $$SS_1 = \sum \frac{(\text{Sum of } each \text{ column})^2}{N \text{ in each column}} - \frac{(\sum X)^2}{N}$$

5. Sum of squares for factor 2 (SS_2):

    Computing formula:

    $$SS_2 = \sum \frac{(\text{Sum of } each \text{ row})^2}{N \text{ in each row}} - \frac{(\sum X)^2}{N}$$

6. Sum of squares for interaction ($SS_{1\times2}$):

    Computing formula:

    $$SS_{1\times2} = SS_B - SS_1 - SS_2$$

2. Degrees of freedom

Total degrees of freedom:

$$(df_T) = N - 1$$

where

$$N = \text{total number of observations}$$

Degrees of freedom within groups:

$$(df_W) = N - k$$

where

$$k = \text{number of cells}$$

Degrees of freedom for factor 1:

$$(df_1) = \text{one less than the number of levels for factor 1}$$

Degrees of freedom for factor 2:

$$(df_2) = \text{one less than the number of levels for factor 2}$$

Degrees of freedom for interaction:

$$(df_{1\times2}) = df_1 \times df_2$$

3. Mean squares

Mean square within groups:

$$(MS_W) = \frac{SS_W}{df_W}$$

Mean square for factor 1:

$$(MS_1) = \frac{SS_1}{df_1}$$

Mean square for factor 2:

$$(MS_2) = \frac{SS_2}{df_2}$$

Mean square for interaction:

$$(MS_{1\times2}) = \frac{SS_{1\times2}}{df_{1\times2}}$$

4. F ratios and tests of significance

Effect of factor 1:

$$F = \frac{MS_1}{MS_W}$$

Effect of factor 2:

$$F = \frac{MS_2}{MS_W}$$

Effect of interaction:

$$F = \frac{MS_{1\times2}}{MS_W}$$

Each computed F value is compared to the critical value from the F table for the degrees of freedom associated with the numerator and denominator *of that test*. If the computed F is less than the table F, H_0 is retained; otherwise, H_0 is rejected in favor of H_1 (the effect is statistically significant).

5. Other considerations

When the F test for a factor is statistically significant, that factor has more than two levels, and the F test for the interaction is *not* significant, the protected t test may be used to ascertain which levels differ from which others. Either null-hypothesis tests or confidence intervals may be used, and the procedures are essentially the same as for the one-way analysis of variance. When the F test for the interaction *is* statistically significant, indicating that the factors operate jointly, the protected t test focuses on differences between *cells* (specific combinations of the two factors). Interaction is an important concept for behavioral science research.

Chapter 17
Chi Square

PREVIEW

Chi Square and Goodness of Fit: One-Variable Problems

What is the difference between *frequency* data and continuous data such as test scores?

What are the procedures for testing the hypothesis that population frequencies are distributed in a specified way?

What is the observed frequency? The expected frequency? The chi square ratio?

What is the correct statistical model to use in this situation?

How do we test chi square for statistical significance?

When should chi square *not* be used?

Chi Square as a Test of Independence: Two-Variable Problems

What are the procedures for testing the significance of the relationship between *two* variables when data are expressed in terms of frequencies?

How do we compute chi square from a 2 × 2 table?

Measures of Strength of Association in Two-Variable Tables

Why is it desirable to convert a statistically significant value of chi square into a measure that shows the strength of the relationship between the two variables?

What are three measures of strength of relationship? When should each of these measures be used?

Summary

Chi Square and Goodness of Fit: One-Variable Problems

Consider the following problems:

1. A publisher introduces three titles for a new literary magazine for women—"Today," "Choice," and "New Alternatives." She wonders whether or not these titles will be equally popular among female consumers. To test the hypothesis of equal preference, she obtains a random sample of 177 women and asks them which *one* of the three titles they like best. She finds that 65 women prefer "Today," 60 prefer "Choice," and 52 prefer "New Alternatives." Are these results sufficient reason to reject the null hypothesis that equal numbers of women *in the population* prefer each of the three titles? Or are these results likely to occur as a result of sampling error if the preferences in the population are in fact equal, in which case the null hypothesis of equality in the population should be retained?

2. A college professor believes that most students in her college would like to eliminate final examinations. She obtains a random sample of students and finds that 160 would like to do away with final exams, while 115 would not. Can she reject the null hypothesis that the issue divides the student population equally?

3. A new proposal favored by students at a particular college requires a two-thirds vote of the faculty for approval. A random sample of 100 faculty members is obtained, and 55 favor the proposal while 45 are opposed. What conclusion should be reached with regard to the null hypothesis that two-thirds of the faculty favor the proposal?

When you use statistics such as t and F, groups are compared in terms of the *amount* of a given characteristic. For example, in Chapter 11, the experimental group given caffeine had greater test scores on the average ($\overline{X} = 81.0$) than the control group not given caffeine ($\overline{X} = 78.0$), and the t test dealt with the average amount of points scored by each group. Not infrequently, however, the data consist only of the *number of objects* (usually persons) *falling in any one of a number of categories,* and information about the amount of the given characteristic is not relevant or not available. That is, the data are in terms of *frequencies* (number of objects). These empirically observed frequencies are compared to frequencies expected on the basis of some hypothesis.

The data in the three problems given at the beginning of this chapter consist of frequencies or head counts—that is, *how many* prefer a particular alternative. The objective is to decide whether or not it is reasonable to conclude that the population frequencies are distributed in a specified way. Thus, in problem 1, the publisher knows how many women in the sample prefer one title to the other two. She would like to know if it is reasonable to conclude that the population from which her sample was randomly drawn is equally divided with regard to preference for the three titles.

The testing of hypotheses about frequency data falling in category-sets is similar in strategy to testing differences between an observed sample mean and the hypothesized value of the population mean. However, a different statistical model must be used with frequency data. The question asked by the investigator dealing with frequencies is: Can the differences between the observed frequencies (symbolized as f_0) and the frequencies expected *if* the null hypothesis is true (symbolized as f_e) be attributed to random sampling fluctuations? Or are the differences likely to be due to nonchance factors — that is, to the falsity of the null hypothesis? The statistic used to test such hypotheses is *chi square* (χ^2), defined by

$$\chi^2 = \sum \frac{(f_0 - f_e)^2}{f_e}$$

where

f_0 = **observed frequency**
f_e = **expected (null-hypothetical) frequency**
$\sum$ **is taken over all the categories**

If the differences between the observed frequencies and the expected frequencies are small, χ^2 will be small. The greater the difference between the observed frequencies and those expected under the null hypothesis, the larger χ^2 will be. If the differences between the observed and expected values are so large collectively as to occur by chance only .05 or less of the time when the null hypothesis is true, the null hypothesis is rejected.

The three problems given at the beginning of this chapter are all concerned with differences in choice among different categories (or levels) of a *single* variable. In the first problem, the variable is literary title preference, and each respondent selects one of three possible choices. In the second case, the variable is attitude toward final examinations, and the student makes one of two possible choices. In problem three, the variable under consideration is attitude toward the new proposal, for which there are also two alternatives.

In all three problems, the expected frequencies must be stated before χ^2 can be determined. Considering problem 1 first, you first set up a table with the three alternatives and tabulate the observed frequencies. (See Table 17.1.) The null hypothesis states that the three titles are equally preferred, so *if* it is true, you would expect the sample of 177 women to be equally divided among the three categories. Thus, the expected frequencies under the null hypothesis are 177/3 or 59, 59, and 59. (Note that the sum of the expected frequencies must be equal to the sum of the observed frequencies.) Is it likely or unlikely that observed frequencies of 65, 60, and 52 would occur if the population frequencies are exactly equal? In other words, are the observed frequencies of 65, 60, and 52 significantly different from the expected frequencies of 59, 59, and 59? To answer this question, χ^2 has been

TABLE 17.1

Chi-square test for literary title choices

Title	Observed Frequency (f_o)	Expected Frequency (f_e)	$f_o - f_e$	$(f_o - f_e)^2$	$\dfrac{(f_o - f_e)^2}{f_e}$
Today	65	59	6	36	.610
Choice	60	59	1	1	.017
New Alternatives	52	59	-7	49	.831

$$\chi^2 = \sum \frac{(f_o - f_e)^2}{f_e} = 1.458$$

computed in Table 17.1. Note that *each* value of $(f_o - f_e)^2$ is divided by its own expected frequency. (All the expected frequencies happen to be equal in this particular problem, but this will not always be true since it depends on the particular null hypothesis.) The resulting three values are then summed to obtain χ^2, which is equal to 1.46.

In order to test the significance of χ^2 using a specified criterion of significance, the obtained value is referred to Table G in the Appendix with the appropriate degrees of freedom. Note that in this table, there is a different value of χ^2 for every *df*. Chi square, like *t*, yields a family of curves, with the shape of a particular curve depending on the *df*. (See Figure 17.1.) For the χ^2 distributions, however, the *df* is based on the number of *categories*, rather than on the sample size as in the case of the *t* distributions. In the one-variable case, the *df* are equal to $k - 1$, where *k* is equal to the number of cate-

FIGURE 17.1

Chi-square distribution for $df = 1$ and $df = 6$

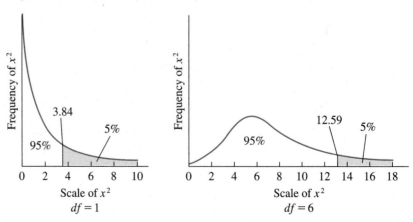

TABLE 17.2

TABLE 17.2

Chi-square test for attitude toward final examinations

Attitude	f_o	f_e	$f_o - f_e$	$(f_o - f_e)^2$	$\dfrac{(f_o - f_e)^2}{f_e}$
For	115	137.5	− 22.5	˙506.25	3.682
Against	160	137.5	22.5	506.25	3.682
	275	275.0			

$$\chi^2 = \sum \frac{(f_o - f_e)^2}{f_e} = 7.364$$

gories of the variable. Thus, in the present problem, $df = 3 - 1 = 2$. For χ^2 to be significant at the .05 level, therefore, the obtained value must be equal to or greater than 5.99. Since the obtained value is only 1.46, the null hypothesis is retained; there is not sufficient reason to reject the null hypothesis that the frequencies in the population are equal. Therefore, we have an insufficient basis for concluding that any particular title or titles is (are) preferred.

The computation of chi square for the second problem posed at the beginning of this chapter is shown in Table 17.2. The null hypothesis specifies that the population of students is equally divided on the issue: 275/2 or 137.5 students will be in favor of final examinations, and 137.5 students will be against final examinations. (Expected frequencies, since they are theoretical, are often in terms of fractions of a person.) The value of χ^2 is equal to 7.36. Since it is larger than the critical value of 3.84 obtained from Table G for $2 - 1 = 1$ df using the .05 criterion of significance, the null hypothesis is rejected; it is *not* reasonable to assume that the student population is evenly divided on this issue. By inspecting the f_0, it may be concluded that a majority is opposed to final examinations.

The analysis of the third problem is shown in Table 17.3. The null hypothesis specifies two-thirds of the faculty in favor of the proposition, so the

TABLE 17.3

Chi-square test for faculty attitudes to new student proposal

Attitude	f_o	f_e	$f_o - f_e$	$(f_o - f_e)^2$	$\dfrac{(f_o - f_e)^2}{f_e}$
For	55	66.67	− 11.67	136.19	2.043
Against	45	33.33	11.67	136.19	4.086
	100	100.00			

$$\chi^2 = \sum \frac{(f_o - f_e)^2}{f_e} = 6.129$$

expected frequency in favor is equal to $2/3 \times 100$ or 66.67. The expected frequency opposed to the proposition is $1/3 \times 100$ or 33.33. The obtained value of χ^2 of 6.13 is statistically significant at the .05 level, since it is greater than the tabled value of 3.84 for $df = 1$. So H_0 is rejected, meaning that it is *not* reasonable to assume that two-thirds of the faculty population is in favor of the proposal. Observing the frequencies, we can conclude that *fewer* than two-thirds are in favor, and that therefore the proposal will not pass. (Lest unwarranted inferences be drawn from Tables 17.2 and 17.3, we hasten to add that these data are entirely fictitious and are intended solely to illustrate the use of χ^2.)

Some Precautions Involving the Use of χ^2

The χ^2 tests described above can only be performed when the observations are *independent*. That is, no response should be related to or dependent upon any other response. For example, it would be incorrect to apply χ^2 with $N = 100$ to the true-false responses of five schizophrenic patients to 20 questionnaire items. The 100 responses are not independent of each other, since the 20 responses given by each patient must be assumed to be mutually related.

Second, any subject must fall in *one* and only one category. Thus, in the problem involving the preference for literary titles, each subject was asked to choose the one title she most preferred.

Third, the computations must be based on all the subjects in the sample. In problem 2, for example, a category of "for" as well as "against" must be included so that χ^2 is based on the total frequency of 275, the total size of the sample. As a check, the sum of the observed frequencies *must* be equal to the sum of the expected frequencies.

One final precaution is concerned with the size of the expected frequencies. Chi square is actually an approximate test for obtaining the probability values for the observed frequencies (that is, the probability of getting the observed frequencies if the null hypothesis is true). This test is based on the expectation that within *any category,* sample frequencies are normally distributed about the population or expected value. Since frequencies cannot be negative, the distribution cannot be normal when the *expected* population values are *close to zero*. The sample frequencies cannot be much below the expected frequency, while they can be much above it—an asymmetric distribution.

Under certain conditions, therefore, you should *not* compute χ^2. For 1 *df*, *expected* frequencies should all be at least 5; otherwise χ^2 should not be used. For 2 *df*, expected frequencies should all exceed 2. With 3 or more *df*, if all expected frequencies but one are greater than or equal to 5 and if the one that is not is at least equal to 1, χ^2 is still a good approximation. In other words, the greater the *df*, the more lenient the requirement for minimum expected frequencies.

Chi Square as a Test of Independence: Two-Variable Problems

In the preceding section, χ^2 was used to test some *a priori* hypothesis about expected frequencies — that is, a hypothesis about f_e values formulated prior to the experiment — which involved frequency data concerning a single variable. Chi square can also be used, however, *to test the significance of the relationship between two variables when data are expressed in terms of frequencies of joint occurrence.*

Suppose you want to find out if men and women differ in their preference for the two major political parties, Democratic and Republican. If the two variables of sex and political party are *not* related (are *independent*), you would expect the proportion of men who prefer Democrats to be the same as the proportion of women who prefer Democrats. (See Table 17.4.) Here, in contrast to one-variable problems, no *a priori* expected frequencies are involved. You are dealing with the relationship between two variables, each of which may have any number of categories or levels. (In this example, each variable has two levels.) The null hypothesis is that the two variables are independent, which (as you will see) implies a set of expected frequencies.

As an illustration, suppose you have a random sample of 108 registered male voters and a random sample of 72 registered female voters. You ask each individual to state which of the two major political parties is preferred. With two categories of sex and two of political parties, there are four possible combined categories or *cells:* male-Democratic, male-Republican, female-Democratic, female-Republican. The results are shown in Table 17.5.

TABLE 17.4

Hypothetical illustration of perfectly independent relationship between sex and political preference, using frequency data

		Sex		Total People
		Female	Male	
Political	Democrats	40	60	100
preference	Republicans	32	48	80
	Total	72 +	108 =	180

55.56% of the women (40/72) prefer Democrats; 44.44% (32/72) prefer Republicans
55.56% of the men (60/108) prefer Democrats; 44.44% (48/108) prefer Republicans
40.00% of the people who prefer Democrats are women (40/100); 60.00% (60/100) are men
40.00% of the people who prefer Republicans are women (32/80); 60.00% (48/80) are men

Thus, there is no relationship between sex and political preference.

TABLE 17.5

Table of frequencies relating sex to political preference (hypothetical data)*

		Sex		Total People
		Female	Male	
Political Preference	Democrats	45 (40)	55 (60)	100
	Republicans	27 (32)	53 (48)	80
	Total	72	+ 108	= 180

Calculations

f_o	f_e	$f_o - f_e$	$(f_o - f_e)^2$	$\dfrac{(f_o - f_e)^2}{f_e}$
45	40	5	25	.625
55	60	− 5	25	.417
27	32	− 5	25	.781
53	48	5	25	.521
				$\chi^2 = 2.344$

* Within each cell, f_o is in the upper left-hand corner and f_e is in parentheses.

Thus, 45 women preferred the Democrats, 55 men preferred the Democrats, 27 women preferred the Republicans, and 53 men preferred the Republicans. The total for the first column (called the *marginal frequency* for that column) indicates that the sample contains 72 (that is, 45 + 27) women. The marginal frequency for the second column shows that the sample contains 108 men. Similarly, the row marginal frequencies indicate that 100 persons of both sexes preferred the Democrats while 80 preferred the Republicans. The total sample size, N, is equal to 180.

The null hypothesis is stated in terms of the independence of the two variables, sex and political preference:

H_0:   sex and political party are independent (*not* related)

H_1:   sex and political party *are* related

The expected frequencies are those of Table 17.4—the frequencies predicated on the independence of the two variables. The procedure for computing the expected frequencies can be summarized as follows: for any cell, the expected value is equal to the product of the two marginal frequencies common to the cell (the row total times the column total) divided by the total N. That is,

$$f_e = \frac{(\text{row total})(\text{column total})}{N}$$

For example, in Table 17.5, the expected frequency for the female-Democratic cell is equal to

$$\frac{(100)(72)}{180} = 40$$

Precisely 72/180 or 40.00% of the sample are women, and 100/180 or 55.56% of the sample are Democrats. So 40.00% of 55.56% or 22.22% of the 180 cases, or $(72/180)(100/180)(180) = 40$ of them, would be expected to be female Democrats if there were no relationship between sex and political preference in the sample. The other f_e values may be found by the same rationale.

The computation of χ^2 is shown in Table 17.5. The resulting value of 2.34 is tested for statistical significance by referring to Table G in the Appendix with the proper df. For a two-variable problem, the df are equal to

$$df = (r - 1)(c - 1)$$

where

$$r = \textbf{number of rows}$$
$$c = \textbf{number of columns}$$

For the 2 × 2 table, the df are equal to $(2 - 1)(2 - 1) = 1$. The reason why there is one degree of freedom in a 2 × 2 table is as follows: Consider the expected frequency of 40 for the female-Democratic cell. Having computed this value, the expected frequency for the male-Democratic cell is fixed (not free to vary). This is because the total of the expected frequencies in the first row of the table, like the total of the observed frequencies, must add up to 100—the marginal frequency for that row. Thus, the expected frequency for the male-Democratic cell is equal to

$$100 - 40 = 60$$

Similarly, the expected values in the first column must add up to the marginal frequency of 72. Having computed the expected value of 40 for the female-Democratic cell, the expected frequency for the female-Republican cell is fixed; it must be

$$72 - 40 = 32$$

Finally, the expected frequency for the male-Republican cell is equal to

$$108 - 60 = 48$$

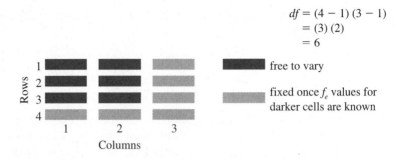

$$df = (4 - 1)(3 - 1)$$
$$= (3)(2)$$
$$= 6$$

FIGURE 17.2
Illustration of degrees of freedom for a 4 × 3 table

Thus, once any one f_e is known, all of the others are automatically determined. In other words, only one f_e is free to vary; the values of the others depend upon its value. Therefore, the 2 × 2 table has one degree of freedom.

In a larger table, all but one of the values f_e in a given row or column are free to vary. Once they are specified, the last one is fixed by virtue of the fact that the expected frequencies must add up to the marginal frequency. An illustration of the degrees of freedom for a 4 × 3 table is shown in Figure 17.2.

Returning to the problem in Table 17.5, the minimum value of χ^2 required to reject H_0 for $df = 1$ and $\alpha = .05$ is 3.84. Since the obtained value of 2.34 is less than the tabled value, you retain the null hypothesis and conclude that there is not sufficient reason to believe that the variables of sex and political preference are related.

The procedures discussed in this section can be used for two-variable problems with any number of levels for each variable. An example of a 3 × 4 problem is shown in Table 17.6. The expected frequency for a given cell is obtained by multiplying the row total for the cell by the column total, and dividing by the total N. For example, f_e for the black-Democratic cell is

$$\frac{(60)(138)}{300} = 27.60$$

The computed value of χ^2 of 27.55 exceeds the tabled value for $(3 - 1)$ $(4 - 1)$ or 6 df of 12.59 at the .05 level of significance. So you reject H_0, and conclude that ethnic group and political preference *are* related.

Computing Formula for 2 × 2 Tables

For a 2 × 2 table, the following formula for χ^2 is equivalent to the one given previously and requires somewhat less work computationally:

OBSERVED FREQUENCIES:

A	B
C	D

$$\chi^2 = \frac{N(AD - BC)^2}{(A + B)(C + D)(A + C)(B + D)}$$

As an illustration, consider the following 2×2 table:

50	70	120
19	41	60
69	111	180 = N

TABLE 17.6

Table of frequencies relating ethnic group to political preference
(hypothetical data)

Ethnic Group	Political Preference				Row Total
	Democratic	Republican	Liberal	Conservative	
Black	36	8	14	2	60
White	84	72	18	26	200
Hispanic	18	8	10	4	40
Column Total	138	88	42	32	300

		Calculations		
f_o	f_e	$f_o - f_e$	$(f_o - f_e)^2$	$\dfrac{(f_o - f_e)^2}{f_e}$
36	27.60	8.40	70.56	2.56
8	17.60	−9.60	92.16	5.24
14	8.40	5.60	31.36	3.73
2	6.40	−4.40	19.36	3.02
84	92.00	−8.00	64.00	.70
72	58.67	13.33	177.69	3.03
18	28.00	−10.00	100.00	3.57
26	21.33	4.67	21.81	1.02
18	18.40	−0.40	.16	0.01
8	11.73	−3.73	13.91	1.19
10	5.60	4.40	19.36	3.46
4	4.27	−0.27	.07	.02
				$\chi^2 = 27.55$

$$\chi^2 = \frac{180(50 \times 41 - 70 \times 19)^2}{(120)(60)(69)(111)}$$

$$= \frac{(180)(720)^2}{(120)(60)(69)(111)}$$

$$= 1.692$$

Measures of Strength of Association in Two-Variable Tables

The χ^2 test of independence allows you to make decisions about *whether* there is a relationship between two variables, using frequency data. Thus, if H_0 is rejected, you conclude that there *is* a statistically significant relationship between the two variables. As we pointed out in Chapter 13, however, statistical significance does not indicate the *strength* of the relationship; a significant result means only that the relationship in the population is unlikely to be zero. Here again, it is very desirable to have a measure of the strength of the relationship—that is, an index of degree of correlation. We will discuss three such measures.

The phi coefficient (ϕ) for 2 × 2 tables. Suppose that a researcher wants to know if there is a relationship between sex and type of preferred living environment. She obtains a sample of 50 men and 50 women, and asks each individual whether he or she would prefer to settle in an urban or suburban area. The results are shown in Table 17.7.

The computed value of χ^2 of 6.895 is statistically significant for $df = 1$ using the .05 criterion. Thus, the two variables of sex and environmental preference *are* related; men tend to prefer urban living and women tend to

TABLE 17.7

Table of frequencies relating sex to type of preferred living environment

		Female	Male		Total People
		\multicolumn{2}{c}{Sex}		
Environmental	Urban	15	28		43
preference	Suburban	35	22		57
	Total	50	+ 50	=	100 = N

$$\chi^2 = \frac{100\,[(15 \times 22) - (28 \times 35)]^2}{(43)\,(57)\,(50)\,(50)}$$

$$= 6.895$$

prefer suburban living. But how strong is this relationship? Since a 2×2 χ^2 table is involved, the answer is provided by converting χ^2 to a correlation coefficient called the *phi coefficient* (symbolized by ϕ), where

$$\phi = \sqrt{\frac{\chi^2}{N}}$$

Thus,

$$\phi = \sqrt{\frac{6.895}{100}}$$
$$= \sqrt{.06895}$$
$$= .26$$

The phi coefficient is interpreted as a Pearson r. In fact, it *is* a Pearson r; you would get the same answer of .26 if you assigned scores of 0 and 1 to sex (for example, 0 = female, 1 = male) and to environmental preference (0 = prefers urban living, 1 = prefers suburban living), and then computed the correlation between the two variables using the usual formula for the Pearson r given in Chapter 12. Thus the absolute value of ϕ varies between 0 and 1. The larger the value of ϕ, the stronger is the relationship between the two variables. When the assignment of score values is arbitrary (as it is in the present case; it would be just as acceptable to score 1 for females and 0 for males, or 1 for prefers urban living and 0 for prefers suburban living), the sign of ϕ is irrelevant, and ϕ is therefore reported without sign. You must look at the data to determine the direction of the findings (in our example, that more men prefer urban living).

In the present example, you would conclude from the value of ϕ that the relationship between sex and environmental preference in the population, while likely to be greater than zero, is also likely to be fairly weak. In other words, there are likely to be numerous exceptions to the conclusion. This is evidenced by the 15 out of 50 women in Table 17.7 who preferred urban living, and the 22 out of 50 men who preferred suburban living. ϕ sums up this information conveniently in a readily interpretable correlation.

Note that there is no need to carry out a special significance test for ϕ. It has already been done with χ^2. In fact, the χ^2 test on a 2×2 table can just as well be viewed as testing the null hypothesis that the population ϕ value equals zero. When χ^2 is large, this hypothesis is rejected, which obviously means that there is some degree of relationship other than zero.

The contingency coefficient, C. When the problem is to determine the strength of the relationship between two variables, data are in terms of frequencies, and a table larger than 2×2 is involved, one can compute a *contingency coefficient* (symbolized by C), where

$$C = \sqrt{\frac{\chi^2}{N + \chi^2}}$$

As an illustration, consider the data in Table 17.6:

$$C = \sqrt{\frac{27.55}{300 + 27.55}}$$

$$= \sqrt{.0841}$$

$$= .29$$

This value of C is necessarily statistically significant (that is, different from zero in the population) because the value of χ^2 on which it is based is statistically significant. But it is fairly small, indicating that the relationship between ethnic group and political preference in the population may well be only a small to moderate one.

Unlike ϕ, however, C is *not* a Pearson r coefficient. Values of C can never be greater than 1.0, but the *maximum* value of the C coefficient in a given situation is strongly influenced by the size of the χ^2 table. It can be shown that the maximum value of C is equal to:

$$C_{max} = \sqrt{\frac{k - 1}{k}}$$

where

k = the *smaller* of r (number of rows) or c (number of columns), and either one if r = c

Thus, for a 3×4 (or 4×3) table, or a 3×3 table,

$$C_{max} = \sqrt{\frac{3 - 1}{3}}$$

$$= .82$$

In spite of this difficulty, C is a long established and widely used index of relationship where larger than 2×2 χ^2 tables are involved.

Cramér's ϕ. An index of strength of relationship for use with larger than 2×2 tables, which is free of the dependence on table size that causes difficulty with the C coefficient, is Cramér's ϕ. This is a generalization of the

2×2 ϕ coefficient that is applicable to $r \times c$ tables of frequencies, and always varies between 0 and 1 regardless of the size of the table. It is simply

$$\text{Cramér's } \phi = \sqrt{\frac{\chi^2}{N(k-1)}}$$

where, as before,

$$k = \text{smaller of } r \text{ or } c \text{ (or, when } r = c, \ k = r = c)$$

For the example in Table 17.6, which is a 3×4 table,

$$\text{Cramér's } \phi = \sqrt{\frac{27.55}{300(3-1)}}$$

$$= \sqrt{.0459}$$

$$= .21$$

Both C and Cramér's ϕ are applicable in the same circumstances. The traditionally used measure is C; Cramér's ϕ is a superior index, but not yet as widely known among behavioral scientists.

Summary

Chi square is used with *frequency* data.

$$\chi^2 = \sum \frac{(f_o - f_e)^2}{f_e}$$

where

$$f_o = \text{observed frequency}$$
$$f_e = \text{expected frequency}$$

1. One-variable problems

$$df = k - 1$$

where

$$k = \text{number of categories of the variable}$$

Expected frequencies are readily determined from the null hypothesis. For example, if H_0 specifies that subjects in the population are equally divided among the k categories, f_e for each category is equal to N/k (where $N = $ number of subjects in the sample).

2. Two-variable problems; test of association

$$df = (r - 1)(c - 1)$$

where

$$r = \textbf{number of rows}$$
$$c = \textbf{number of columns}$$

For a cell in a given row and column, the expected frequency is equal to

$$f_e = \frac{\textbf{(row total)(column total)}}{N}$$

For a 2×2 table, the following formula for χ^2 is equivalent to determining expected frequencies and using the usual formula, and requires somewhat less work computationally:

A	B
C	D

$$\chi^2 = \frac{N(AD - BC)^2}{(A + B)(C + D)(A + C)(B + D)}$$

3. Measures of strength of association in two-variable tables

2×2 tables:
  Compute the phi coefficient:

$$\phi = \sqrt{\frac{\chi^2}{N}}$$

ϕ is interpreted as a Pearson r.

Larger tables:
  Compute Cramér's ϕ:

$$\textbf{Cramér's } \phi = \sqrt{\frac{\chi^2}{N(k - 1)}}$$

where

$k = $ the *smaller* of r (number of rows) or c (number of columns);
$k = $ either one if $r = c$

These two measures are statistically significant if χ^2 is statistically significant.

4. *Some precautions on the use of χ^2*

χ^2 should be used only when the observations are independent—that is, when no observation is related to or dependent upon any other observation. Do *not* compute χ^2 under any of the following conditions:

a. $df = 1$, and any *expected* frequency is less than 5.
b. $df = 2$, and any *expected* frequency is less than 3.

Chapter 18

Nonparametric and Distribution-Free Methods

PREVIEW

Introduction

What is a parametric statistical test? A nonparametric statistical test? A distribution-free statistical test?

Basic Considerations

What are the advantages of using nonparametric and distribution-free tests?

What are the disadvantages of using nonparametric and distribution-free tests?

What are some important considerations about significance tests based on ranks?

The Difference Between the Locations of Two Independent Samples: The Rank-Sum Test

What are the procedures for testing the hypothesis that two independent samples of ranked data come from populations with equal locations?

What is the power efficiency of the rank-sum test?

When should the rank-sum test be used in place of the t test for two independent sample means?

How do we convert a significant result to a measure of strength of relationship?

Differences Among the Locations of Two or More Independent Samples: The Kruskal-Wallis H Test

What are the procedures for testing the hypothesis that three or more independent samples of ranked data come from populations with equal locations?

What is the power efficiency of the Kruskal-Wallis H test?

When should the Kruskal-Wallis H test be used in place of the one-way analysis of variance?

What is the protected rank-sum test?

How do we convert a significant result to a measure of strength of relationship?

(continued next page)

Introduction

A *parametric* statistical test involves the estimation of the value of at least one population parameter. For example, the within-group sample variance calculated in the t and F tests is an estimate of the corresponding within-population variance. A *nonparametric* test, on the other hand, does not require such estimation. Also, the mathematical derivations of the t and F tests assume that each population being randomly sampled has a *normal* distribution. A *distribution-free* statistical test, by contrast, requires no assumption about the shape of the distribution in the population. While the terms "nonparametric" and "distribution-free" have different meanings, most statistical tests that meet one of these conditions also satisfy the other, so the terms tend to be used interchangeably in practice.

A prominent example of such methods is the chi-square test applied to one-way and two-variable problems. This test is clearly distribution-free; in fact, it tests the equality of entire distributions of frequencies. (It is also non-parametric, but the argument for this is complicated and technical.) It is treated in a separate chapter in this and most textbooks because of its wide applicability, and because there are no practical alternative procedures that might be used in its place.

Basic Considerations

Advantages

In this chapter, we will consider some tests that are parallel to t and F tests, but that are both nonparametric and distribution-free. Why should one wish to employ such tests?

The obvious advantage of these tests is that they do not require the population being sampled to be normally distributed. When we discussed the assumption of normality in connection with t and F tests, we stated that these tests were robust; moderate departures from normality of distributions did not seriously affect their validity. But it is not at all rare in the behavioral sciences to encounter data that are grossly nonnormally distributed. Consider, for example, a typical distribution of reaction times: the bulk of the reaction times will be fairly close to zero, a fair number will be relatively long, but none can be less than zero. Thus the distribution is quite asymmetrical, and therefore necessarily substantially *non*normal. So when we have reason to believe that some population distributions depart greatly from normality, distribution-free methods (which require no assumption about population distribution shape) are desirable. These methods do generally assume, however, that the populations being compared have the *same* shape and variability.

A second advantage that has been claimed for these tests is that they require less computation than the parametric methods. Thus they may be used to provide quick and generally good approximations to the results that would be obtained by testing comparable null hypotheses using the parametric t and F tests. However, now that computers and electronic calculators have become much more widely available, this advantage has lessened considerably. You pay a price for using distribution-free methods when parametric methods are suitable, as we will see.

Disadvantages

It may have occurred to you to ask, "Why *ever* use parametric tests, which make normality assumptions about unobservable population distributions, when distribution-free tests are available? After all, one can never *know* that the shape of a population distribution is normal." The answer is that when populations are normally distributed, the probability that a distribution-free test will reject a false null hypothesis (its statistical power) is *smaller* than for the analogous parametric test. That is, under similar conditions, nonparametric tests are more likely to lead to a Type II error if the populations are normally distributed.

It therefore follows that when populations are normally distributed, the sample size required by the parametric test (N_p) is smaller than that required by the distribution-free test (N_d) in order to obtain the same amount of power. The ratio of these sample sizes expressed as a percent, or $(N_p/N_d) \times 100\%$, is called the *power efficiency* of the distribution-free or nonparametric test. Suppose that for a given difference between population means, a given significance criterion, α, and a specified power, a certain distribution-free test requires a total sample of $N_d = 80$ cases. Its parametric alternative, however, requires only $N_p = 72$ cases. The power efficiency of the distribution-free test would be $72/80 = 90\%$, so the parametric test requires only 90% as many cases to have the same probability of rejecting the null hypotheses.

When populations are normally distributed, the power efficiencies of distribution-free tests are almost always less than 100%, and for some tests, much less. This means that they require more cases in order to have the same power. In other words, they have less power for the same number of cases. Therefore, using a distribution-free test is likely to involve a rather unpleasant choice: either you must expend the effort to obtain more cases in order to have as much power as the corresponding parametric test, or you must run a greater risk of making a Type II error.

For these reasons, you should *not* rush to use distribution-free tests unless substantial nonnormality is believed to exist in the population. This is especially true when you are using two-tailed tests and when the sample sizes are *not* very small (say, 20 or more). But if you do have good reason to believe that a population distribution departs substantially from normality, or if you wish to be particularly conservative on the side of Type I errors, an appropriate nonparametric/distribution-free test should be chosen.

The Nature of Significance Tests Based on Ranks

We have already considered one procedure based on ranks: the Spearman rank-order correlation coefficient (r_s) in Chapter 13. It is used when the standings of subjects on two variables are expressed in ranks rather than in scores. r_s enables you to measure the degree of relationship between the two variables, and to test it for statistical significance. Actually r_s is a nonparametric measure of correlation, and its significance test is distribution-free.

Among the most important nonparametric and distribution-free tests are those that deal with differences in *location* between populations, based on ranks. Recall from Chapter 4 that "location" is a general term describing where a distribution falls, often called its "central tendency." In parametric statistics, the location of a population or sample is customarily indexed by its mean. When we conclude from a t test that scores in population 1 are larger than those in population 2, we mean literally that $\mu_1 > \mu_2$. But in nonparametric tests, the same statement has a different meaning: if we draw at random a case from population 1 and another from population 2 and compare them, and do this repeatedly, more than half the time the case from population 1 will be larger than the case from population 2 (technically, population 1 is "stochastically larger" than population 2). This can be taken for all intents and purposes as an assertion that $Mdn_1 > Mdn_2$.

Significance tests based on ranks are used when the original data come in the form of ranks (a relatively rare occurrence), or after a set of scores has been converted into ranks for the purpose of performing the test. In converting a set of N scores to ranks, the same procedure is used as with r_s: the rank 1 is assigned to the smallest observation and the rank N to the largest. If ties occur, the mean of the ranks in question is assigned to each of the tied scores.

We will consider three different statistical tests based on ranks: a test of the difference in location between two independent samples, a test of the differences in location among k (more than two) independent samples, and a test of the difference in location between two matched (dependent) samples.

What makes tests based on ranks work so simply is the fact that the data for any problems involving a total of N cases (where there are no ties) are the positive integers 1, 2, 3, . . . , N. Thus statistical functions of these data are simple functions of N. For example, it can easily be proved that the *sum* of the first N integers (ranks 1 to N) is as follows:

$$\sum R = 1 + 2 + 3 + \cdots + N$$

$$= \frac{N(N + 1)}{2}$$

Suppose that there are a total of 22 cases. Whatever their original raw scores may be, the sum of their *ranks* must be $(22)(23)/2 = 253$. And since the mean is equal to the sum divided by N, it therefore follows that the mean of a complete set of ranks (symbolized by $\bar{R}$) is equal to

$$\bar{R} = \frac{\sum R}{N}$$

$$= \frac{N(N + 1)}{2N}$$

$$= \frac{N + 1}{2}$$

Thus, the mean rank for 22 observations is simply $23/2 = 11.5$. The mean rank can be thought of as the "expected" rank of a case drawn at random from the N ranks, in the sense that the mean is the best guess as to the value of this case. Therefore, if we draw two cases at random when $N = 22$, the expected sum of the ranks (symbolized by T_E) is $2(11.5) = 23$. For three cases, it is $3(11.5)$ or 34.5. Generally, if we draw a subset of N_1 cases at random from the total N, their expected sum of ranks is N_1 times the mean rank:

$$T_E = N_1 \overline{R}$$

$$= \frac{N_1(N + 1)}{2}$$

It can also be shown that if N_1 cases are randomly and repeatedly drawn from the complete set of N ranks, and the sum of the ranks (T) is obtained for each sample, the standard deviation of the resulting sampling distribution of T values (or the *standard error* of T) would be

$$\sigma_T = \sqrt{\frac{N_1 N_2 (N + 1)}{12}}.$$

where

$$\sigma_T = \text{standard error of } T$$
$$N_2 = N - N_1$$

The value of 12 in the denominator is a constant, and is not affected by any of the N values. Regardless of the shape of the population distribution of raw scores, this sampling distribution of a sum of a subset of ranks is approximately normal in form. This is *not* an assumption, but a provable mathematical necessity.

There now exist all the ingredients for a statistical test: a sample statistic (T) with a known (in this case, normal) sampling distribution, whose standard deviation is known (σ_T), and a null-hypothetical value (T_E). This leads us into the test of the difference in location between two independent samples. We will subsequently consider two other important applications of significance tests based on ranks: a test of the differences in location among more than two independent samples, and a test of the difference in location between two matched (dependent) samples.

The Difference Between the Locations of Two Independent Samples: The Rank-Sum Test

Rationale and Computational Procedures

The rank-sum procedure is used to test the null hypothesis that two independent samples of ranked data come from populations with equal locations. Table 18.1 gives scores for a total of $N = 22$ subjects, of whom $N_1 = 10$ are in Group 1 and $N_2 = 12$ are in Group 2. The normality assumption appears to be quite dubious, so a distribution-free test is appropriate.

TABLE 18.1

The rank-sum test for two independent samples

Group 1			Group 2	
X	R		X	R
55	3		52	1
62	7		53	2
69	11		56	4
78	15		58	5
79	16		60	6
84	18		63	8
92	19		64	9
96	20		65	10
98	21		71	12
99	22		74	13
			76	14
			81	17

$$T_1 = 152 \qquad\qquad T_2 = 101$$
$$N_1 = 10 \qquad\qquad N_2 = 12$$
$$\bar{R}_1 = \frac{152}{10} \qquad\qquad \bar{R}_2 = \frac{101}{12}$$
$$= 15.20 \qquad\qquad = 8.42$$

$$N = N_1 + N_2 = 22$$
$$\sum R = N(N + 1)/2 = (22)(23)/2 = 253$$
$$= T_1 + T_2 = 152 + 101 = 253$$

First, all 22 observations are subjected to a single ranking from the lowest ($X = 52$), which is assigned a rank of 1, to the highest ($X = 99$), which is assigned a rank of 22. The X scores are ignored from this point on, and the analysis proceeds solely with the ranks.

The question now is whether the ranks in Group 1 (or Group 2) are generally larger or smaller than one would expect by chance. The sum of the ten ranks in Group 1, symbolized by T_1, is 152. According to the null hypothesis, the expected sum of the ranks is

$$T_E = \frac{N_1(N + 1)}{2}$$

$$= \frac{(10)(23)}{2}$$

$$= 115$$

Is the observed T_1 of 152 a statistically significant departure from the null-hypothetical (expected) T_E value of 115? The analysis is quite similar to the normal curve procedures for tests about the mean of one population (Chapter 10):

$$z = \frac{T_1 - T_E}{\sigma_T}$$

$$= \frac{T_1 - T_E}{\sqrt{\dfrac{N_1 N_2 (N + 1)}{12}}}$$

Instead of comparing a sample mean $(\overline{X})$ to a null-hypothesized population value (μ) and dividing by the standard error of the mean, as in Chapter 10, you compare the sample rank sum (T_1) to the null-hypothesized expected value (T_E) and divide by the standard error of T. (Remember that the value of 12 in the denominator is a constant, and has no relation to any of the N values.) And since the sampling distribution of T is approximately normal, you can determine whether the observed T_1 departs significantly from T_E by using the normal curve model and z.

For the data in Table 18.1, the results are:

$$z = \frac{152 - 115}{\sqrt{\dfrac{(10)(12)(23)}{12}}}$$

$$= \frac{37}{15.17}$$

$$= 2.44$$

So the 37 rank-sum units by which Group 1 deviates from T_E place it 2.44 standard error units above the mean of a normal distribution of rank-sums. The computed z value of 2.44 exceeds the critical z value of ± 1.96 (for $\alpha = .05$, two-tailed). So you reject the null hypothesis, and conclude that the ranks in Group 1 are significantly higher than chance expectation. This in turn implies that the ranks in Group 2 are significantly lower than chance expectation, and therefore that the ranks in Group 1 are generally higher than those in Group 2. You can therefore conclude that the location of the two populations from which these samples were randomly drawn differs on X, with the first population being stochastically higher than the second. A more comprehensible way to say this is that the median of the first population is greater than the median of the second population.

It makes no difference which of the two sets of ranks you use to compute T and T_E. For Group 2, $T_2 = 101$, and T_E would now be equal to $(12)(23)/2$ or 138. The value of σ_T is unchanged, so z would equal $(101 - 138)/15.17$ or -2.44. And you would reach exactly the same conclusion, with the minus sign indicating that the group in question (Group 2) is the one with the lower location. Although you need compute only T_1 (or T_2) to carry out the statistical test, it is desirable to calculate both values (and verify that

$T_1 + T_2 = N[N + 1]/2$ as a check against errors in ranking and/or calculating.

The rank-sum test for two samples depends on the fact that the sampling distribution of T is well approximated by the normal curve. This will be true provided that the samples are not too small, say at least 6 to 8 cases in each. With fewer than that, a different approach is needed, and tables of the small-sample exact distribution values of T (or of the closely related Mann-Whitney U statistic) are then required. This method is not included here because it is relatively rare that differences in location are large enough to be detectable with such small samples. Other statistical refinements, such as a correction for continuity that slightly reduces the absolute size of the numerator of the z formula and a procedure for reducing σ_T to take tied ranks into account, are omitted because their effects are small and tend to cancel out.

The parametric analog of the rank-sum test is the t test for two independent sample means (Chapter 11). Relative to the t test, the rank-sum test has power efficiency ranging from 92% for small samples to 95% for large samples. For a given set of conditions, when populations are normally distributed, the same power is obtained using the t test with 92–95% of a sample size as is required by the rank-sum test. Compared to other nonparametric tests, those based on ranks have fairly high power efficiencies, so they should be used when the normality assumption is in serious doubt. When this is not the case, the behavioral scientist usually cannot afford the luxury of what amounts to throwing away 5–8% of the cases.

Measure of Strength of Relationship: The Glass Rank Biserial Correlation

Throughout this book we have stressed the general insufficiency of merely computing a test of statistical significance, which indicates only whether or not there is some (nonzero) effect in the population, and the desirability of indexing the *strength* of an observed relationship. For example, in the case of the parametric t test for two independent sample means, the point biserial r served as such a measure (Chapter 13). The point biserial r is the product moment (Pearson) correlation between the dichotomous variable, membership in Group 1 or Group 2, and the continuous variable under study.

An analogous measure of the strength of the relationship between the group membership dichotomy and the *rank* values for two groups is provided by the Glass rank biserial correlation coefficient, symbolized by r_G. It is computed using the following formula:

$$r_G = \frac{2\,(\bar{R}_1 - \bar{R}_2)}{N}$$

where

$$\overline{R}_1 = \text{mean of ranks in Group 1}$$
$$\overline{R}_2 = \text{mean of ranks in Group 2}$$
$$N = \text{total number of observations}$$

The r_G coefficient is *not* a product moment (Pearson) r computed on the rank values. However, it has the same limits: it may take on values from -1 to $+1$, with its sign depending on which group is called Group 1, and the magnitude is interpreted much like that of any other correlation coefficient. The maximum value of $+1$ or -1 occurs when there is no overlap in ranks between the two groups, with one group containing all the highest ranks while the other group contains all the lowest ranks. (By contrast, r_{pb} equals $+1$ or -1 only when the values within each group are identical, that is, when the variance within each group is zero. This is clearly a higher standard than nonoverlap between groups.)

For the data in Table 18.1, $\overline{R}_1 = 15.20$ and $\overline{R}_2 = 8.42$, and $N = 22$. Thus,

$$r_G = \frac{2(15.20 - 8.42)}{22}$$

$$= .62$$

This indicates a fairly high degree of relationship between group membership and ranking on X.

Be careful not to confuse r_G with the Spearman rank-order correlation coefficient r_s. The Spearman r_s coefficient gives the relationship between two continuous variables, each of which has been independently ranked. The r_G procedure, on the other hand, provides a measure of the relationship between the dichotomous variable of group membership and a *single* continuous variable that has been ranked across all N observations. The r_s coefficient is thus the direct analog to the general Pearson r, while r_G is the direct analog to r_{pb}.

Differences Among the Locations of Two or More Independent Samples: The Kruskal-Wallis H Test

Rationale and Computational Procedures

When the locations of more than two independent samples are to be compared, a distribution-free test based on ranks is provided by the Kruskal–Wallis H test. This procedure is analogous to the parametric F test of one-way analysis of variance, and it uses the same logic. However, because statistics based on ranks are simple functions of N, the computation of H is much easier than is the computation of F. In fact, H is a simple function of just the sum of squares between groups (SS_B) exactly as defined in Chapter

15, except that SS_B is computed on the ranks instead of the scores. Also, just as the t test is a special case of one-way ANOVA where there are only two groups, the rank-sum test described previously is a special case of the Kruskal-Wallis H test where there are only two groups. That is, when k (the number of groups) is equal to 2, the rank-sum test and the Kruskal-Wallis H test give identical results.

By way of illustration, the data from Table 15.3 have been converted into ranks. (See Table 18.2.) The total of $N = 27$ observations have been ranked from the lowest (rank = 1) to the highest (rank = 27) and organized into $k = 3$ groups, in exactly the same order as they appeared in Table 15.3. Tied scores were assigned mean ranks in the usual manner. For example, there are two scores of 9 that should occupy ranks 5 and 6, and each has been assigned the rank of 5.5 (the mean of 5 and 6). The next highest score, 10 (which is not tied), receives a rank of 7.

The sum of the ranks for each of the k groups is now obtained; $T_1 = 113.5$, $T_2 = 189.5$, and $T_3 = 75$. (To guard against errors in ranking or summing, the sum of all T values should be verified as equal to $\sum R$ or $N[N + 1]/2$, as shown at the bottom of the table.) The next step is to compute the sum of squares between groups, SS_B, for the ranks. The computing formula given in Chapter 15 may be used, or the following (somewhat simpler) one that capitalizes on the fact that the data are in rank form:

$$SS_B = \frac{T_1^2}{N_1} + \frac{T_2^2}{N_2} + \cdots + \frac{T_k^2}{N_k} - \frac{N(N + 1)^2}{4}$$

TABLE 18.2

The Kruskal-Wallis H test for $k = 3$ independent samples

Group 1		Group 2		Group 3	
X	R	X	R	X	R
15	19	17	21	6	2
18	22	22	27	9	5.5
12	12	5	1	12	12
12	12	15	19	11	8.5
9	5.5	12	12	11	8.5
10	7	20	24	8	4
12	12	14	16.5	13	15
20	24	15	19	14	16.5
		20	24	7	3
		21	26		
$T_1 = 113.5$		$T_2 = 189.5$		$T_3 = 75$	
$N_1 = 8$		$N_2 = 10$		$N_3 = 9$	

$$N = N_1 + N_2 + N_3 = 27$$
$$\sum R = N(N + 1)/2 = (27)(28)/2 = 378$$
$$= T_1 + T_2 + T_3 = 113.5 + 189.5 + 75 = 378$$

Substituting the values from Table 18.2 gives

$$SS_B = \frac{(113.5)^2}{8} + \frac{(189.5)^2}{10} + \frac{(75)^2}{9} - \frac{27(28)^2}{4}$$

$$= 5826.31 - 5292.00$$

$$= 534.31$$

The final step is to compute the Kruskal-Wallis H statistic:

$$H = \frac{12 SS_B}{N(N + 1)}$$

The value of 12 in the numerator is a constant, and arises from the fact that the variance of a full set of N untied ranks is equal to $(N + 1)^2/12$. In our example,

$$H = \frac{(12)(534.31)}{(27)(28)}$$

$$= 8.48$$

Under the null hypothesis that the locations of the k populations are identical, H is distributed approximately as χ^2 with $k - 1$ degrees of freedom. Thus the computed H value of 8.48 is referred to Table G of the Appendix, and is found to be larger than the value of 5.99 that corresponds to the .05 criterion of significance for 2 df. You should therefore reject the null hypothesis that the three populations have equal locations.

The chi-square approximation used to test H for statistical significance is a good one unless the size of any group is very small. If there are no more than three groups, each group should have at least five cases in it; otherwise the Kruskal-Wallis H test should not be used. With more than three groups, as few as two cases in a group are sufficient. The existence of tied ranks tends to make the H test conservative (smaller and less likely to yield statistical significance), but this tendency is slight unless ties are very long and numerous. It is possible to correct for ties, and to perform this test for very small samples. Finally, as compared to the analogous parametric F test in a one-way ANOVA, the H test has power efficiency ranging from 90% for small samples to 95% for large samples. The same comments apply here as for the rank-sum test.

Multiple Comparisons: The Protected Rank-Sum Test

We have seen that when the overall F test in an analysis of variance is statistically significant, an additional procedure is necessary in order to determine which populations differ significantly. The protected t test (Fisher's LSD method) is used for this purpose. Similarly, if and only if a Kruskal-Wallis H is statistically significant, a method called the protected rank-sum test may be used to determine which pairs of populations differ significantly in location. As with the protected t test, you may perform any or all of the $k(k-1)/2$ pairwise rank-sum tests for independent samples; the tests are protected against a large experimentwise Type I error rate by the precondition that H be significant.

To illustrate, let us return once again to the data in Table 18.2. Since H was statistically significant, any or all of the three pairs of groups (1 versus 2, 1 versus 3, or 2 versus 3) may be compared by using the protected rank-sum test. In order to do so, however, *the scores must be reranked for each test.* For example, if Groups 1 and 2 are being compared, the third group is ignored. The third score in Group 2 is ranked 1 (the lowest score), the fifth score in Group 1 is now ranked 2, the score right below this one is now ranked 3, and so on until the highest score in these two groups (the second score in Group 2) receives a rank of 18. Thus $N_1 = 8$, $N_2 = 10$, and $N = 18$ for this comparison. Upon reranking it is found that $T_1 = 59.5$ (and $T_2 = 111.5$), $T_E = (8)(19)/2 = 76$, and $\sigma_T = \sqrt{(8)(10)(19)/12} = 11.25$. Therefore, using the formula for the rank-sum test given previously, we have

$$z = \frac{T_1 - T_E}{\sigma_T}$$

$$= \frac{59.5 - 76}{11.25}$$

$$= -1.47$$

This z value is not statistically significant. When Group 1 and Group 3 are compared, a nonsignificant z value of 1.73 is obtained. But when Group 2 and Group 3 are compared, a statistically significant z value of 2.69 results. Therefore the conclusion is that the locations of these three populations are not the same because the location (or median) of population 2 is greater than that of population 3. However, the data do not warrant concluding that the location of population 1 differs from either of the other two.

Measure of Strength of Relationship: ε_R

A measure of the strength of the relationship between membership in one of the k groups and rank on the dependent variable may be obtained by computing epsilon. This is the same index that was used in connection with one-way ANOVA, but applied to ranks rather than scores. Just as ε for scores was a function of F and dfs, so ε_R may be found from H, k, and N:

$$\varepsilon_R = \sqrt{\frac{H - k + 1}{N - k}}$$

where

$$k = \text{number of groups}$$
$$N = \text{total number of observations}$$

For example, the ranked attitude scores in Table 18.2 yielded an H of 8.48. Therefore, for these data,

$$\varepsilon_R = \sqrt{\frac{(8.48 - 3) + 1}{27 - 3}}$$

$$= \sqrt{\frac{6.48}{24}}$$

$$= \sqrt{.27}$$

$$= .52$$

For the continuous score data in Table 15.3, ε was found to be equal to .47 — a quite similar value. This similarity will usually be the case. Thus, by either route we find that the observed relationship in the sample is fairly strong.

Just as the statistical significance of ε is automatically given by F, so the significance of ε_R is given automatically by H. So as with the other measures of strength of relationship that we have discussed, there is no need to conduct an additional significance test for ε_R.

The Difference Between the Locations of Two Matched Samples: The Wilcoxon Test

Rationale and Computational Procedures

We saw in Chapter 11 that when a comparison is to be made between the means of two matched or dependent samples, a different procedure is used than when the samples are independent. For each of the N matched pairs, the difference (D) between the X_1 and X_2 values is found, and the null hypothesis tested is that the population mean of the D values is zero. Rejection of this null hypothesis necessarily implies that μ_1 and μ_2 are not equal.

The same distinction applies for the appropriate rank-based test of the difference in location between two samples. The rank-sum test discussed earlier is used with independent samples, while a different technique (called the Wilcoxon matched-pairs signed ranks test) is used with dependent samples. The Wilcoxon test also uses $D = X_1 - X_2$ values, but considers only the ranks of the absolute values of the scores.

As an example, let us consider an investigation into the question of whether sensitivity training improves extrasensory perception (ESP) scores. A sample of 12 subjects is tested for ESP before (X_2) and again after (X_1) a

TABLE 18.3

The Wilcoxon test for two matched samples

Subject	X_1	X_2	D $(= X_1 - X_2)$	$R_{|D|}$	$R(+)$	$R(-)$
1	33	38	− 5	7		7
2	45	43	+ 2	2	2	
3	50	42	+ 8	10	10	
4	45	44	+ 1	1	1	
5	46	49	− 3	3.5		3.5
6	45	41	+ 4	5.5	5.5	
7	28	22	+ 6	8	8	
8	43	46	− 3	3.5		3.5
9	32	32	0			
10	40	31	+ 9	11	11	
11	34	27	+ 7	9	9	
12	40	44	− 4	5.5		5.5

$$\sum R = 66 \qquad T_1 = 46.5 \qquad T_2 = 19.5$$
$$(= T_1 + T_2)$$

$N = 11$ (deleting subject 9 for whom $D = 0$)

$$\sum R = N(N + 1)/2 = (11)(12)/2 = 66$$

series of ten training sessions, and the results are shown in Table 18.3. In this experiment, therefore, the samples are dependent because the two observations in each pair come from the same subject. (It should be noted, however, that this matched design is obtained whenever there is some connection between the observations making up a pair. More exactly, it holds whenever there is a nonzero population correlation between the paired observations.)

The first two columns in Table 18.3 give the posttraining (X_1) and pretraining (X_2) ESP scores for the twelve subjects. The third column shows each subject's change or difference score, where $D = X_1 - X_2$. The D values are positive when the ESP score has increased following sensitivity training, and negative when it has decreased. As was the case in the parametric matched t test, all further analyses focus on these D scores.

First, all cases where $D = 0$ are dropped. Case 9 in the table is therefore deleted, and the analysis proceeds on the basis of $N = 11$. Then, *ignoring the signs* of the D values, they are rank ordered from the smallest (rank = 1) to the largest (rank = N), with ties being resolved in the usual way. This has been done in column $R_{|D|}$. This results in the familiar set of ranks from 1 to N whose statistical properties are known; for example, their sum must equal $N(N + 1)/2$, which for $N = 11$ is equal to $(11)(12)/2$ or 66.

If the population locations of X_1 and X_2 are the same, then (ignoring pairs where $D = 0$) the sum of the ranks of the absolute D values should be approximately equally divided between those that are positive ($R+$) and those that are negative ($R-$). That is, there should be an equal number of cases where $X_1 > X_2$ and where $X_1 < X_2$. Since the sum of *all* the ranks equals $N(N + 1)/2$, the expectation under the null hypothesis is that the sum of the ranks for the positive Ds and the sum of the ranks for the negative Ds will *each* equal half the sum. That is, the expected result under the null hypothesis is

$$T_E = \frac{1}{2} \frac{N(N + 1)}{2}$$

$$= \frac{N(N + 1)}{4}$$

For the data in Table 18.3,

$$T_E = \frac{(11)(12)}{4}$$

$$= 33$$

And this is half of 66, the sum of all the ranks.

In the columns headed $R(+)$ and $R(-)$ in Table 18.3, the $R_{|D|}$ values for positive and negative Ds have been segregated. The sum of the ranks of the positive D values, T_1, is equal to 46.5. The sum of the ranks of the negative D values, T_2, is equal to 19.5. Except as a check, only one of these values is needed, so let us choose $T_1 = 46.5$. This value departs 13.5 rank-sum units from the value expected under the null hypothesis, $T_E = 33$. You are then left with the question of whether or not this departure is "sufficiently unlikely" to reject the null hypothesis, using (say) the $\alpha = .05$ decision rule.

A method for answering this question is readily at hand. In repeated random sampling, T_1 (and also T_2) is approximately normally distributed when the null hypothesis is true. The standard deviation of the sampling distribution of this matched-pairs signed ranks T (the standard error), symbolized by σ_{T_M}, is

$$\sigma_{T_M} = \sqrt{\frac{(2N + 1)T_E}{6}}$$

Thus we have a sampling distribution that is approximately normal, whose mean is T_E and whose standard deviation is σ_{T_M}, and the null hypothesis of equal positive and negative rank sums. This gives us all the ingredients for a significance test using the normal curve model:

$$z = \frac{T_1 - T_E}{\sigma_{T_M}}$$

$$= \frac{T_1 - T_E}{\sqrt{\frac{(2N + 1)T_E}{6}}}$$

For data in Table 18.3, we found that the null hypothesis leads to an expected value of T_1 (and T_2) of $T_E = 33$. The observed $T_1 = 46.5$. Therefore,

$$z = \frac{46.5 - 33}{\sqrt{\dfrac{[(2)(11) + 1](33)}{6}}}$$

$$= \frac{13.5}{\sqrt{\dfrac{(23)(33)}{6}}}$$

$$= 1.20$$

Thus the observed departure of 13.5 rank-sum units from the null-hypothetical expected value of 33 is only a little more than one standard error, and is far less than the value of 1.96 needed to reject H_0, using the two-tailed .05 criterion of significance. Therefore, the data do *not* justify the conclusion that ten sessions of sensitivity training will improve ESP scores in the population from which this sample was randomly drawn. The results do not permit the conclusion that no effect exists, of course, since the probability of a Type II error has not been determined. Only if the population effect were very large would there be a reasonably good chance of rejecting the null hypothesis with such a small sample.

Here again, it does not matter whether T_1 or T_2 is used in the statistical test, since both are necessarily at equal distances from T_E. The only effect of replacing T_1 with T_2 is to change the sign of z.

Relative to the matched t test, the Wilcoxon test has power efficiency ranging from 92% for small samples to 95% for large samples. This means that if the parametric assumptions of the t test for matched samples were valid, our choice of the Wilcoxon test would result in somewhat lower power than the t test would have; our $N = 11$ for the Wilcoxon test is about as powerful as a t test using $N = (.92)(11) = 10$ pairs. (A t test applied to the 12 D values is also not significant; it equals 1.28, very close to the z of 1.20.)

The Wilcoxon test should not be used if the sample size is smaller than about 8, since the approximation to normality will then not be accurate. This is unlikely to be a serious hardship, however, since the power of any statistical test that is based on so paltry an amount of information will usually be quite poor.

Measure of Strength of Relationship: The Matched-Pairs Rank Biserial Correlation

The matched-pairs rank biserial correlation, symbolized by r_C, expresses the strength of the relationship between condition (such as posttest versus pretest) and the dependent variable. Another way to describe r_C is that it indexes the degree of relationship between the sign of D and its rank. It is computed as follows:

$$r_C = \frac{4(T_1 - T_E)}{N(N + 1)}$$

The r_C index can take on values from -1 to $+1$. It is equal to zero when T_1 does not differ from the null-hypothetical value T_E, and equals ± 1 when all D values have the same sign. It is not a product-moment correlation coefficient, however, but is part of the same system as r_G. The sign of r_C indicates whether the positive or negative sums of ranks is larger.

In Table 18.3, the Wilcoxon test was not statistically significant. Therefore, there is little reason to compute r_C for these data. As a guideline that you may use when you do have statistically significant results that you wish to convert to r_C, however, we will illustrate the calculation of r_C for these data:

$$r_C = \frac{4(46.5 - 33)}{(11)(12)}$$

$$= .41$$

This value cannot be statistically significant, since the Wilcoxon test for the same data did not yield significance. It may be used to describe the relationship within this particular sample, but there is insufficient reason to conclude that the corresponding population correlation is different from zero.

Methods for Approximating Statistical Significance: Tests Based on Medians and Signs

Since adjacent scores may be at varying distances from each other, while ranks are always one unit apart, measurement information is lost when scores are transformed to ranks. For example, suppose that the three lowest IQs of 72, 83, and 85 in a particular sample are converted into ranks of 1, 2, and 3. It is no longer possible to tell that the lowest IQ was well below the other two, or that the IQs ranked second and third were quite close in value. Thus, a considerable loss of the original interval information has occurred. The statistical tests discussed previously compensate for this loss of information by being distribution-free, and by suffering only a modest loss of power.

There are other nonparametric and distribution-free statistical tests that are fairly common, but that give up even more information than is lost when going from scores to ranks. The only information used in these tests is whether an observation falls above some point in a distribution. This greatly simplifies the data and computations, but a substantial price is paid in power efficiency. Unlike the tests based on ranks, whose power efficiencies range between 90% and 95%, the power efficiencies of these tests range from 64% to 95% and are in practice usually between 65% and 70%.* So using these tests when a parametric test (t or F test) would have been suitable has about the same effect on power as randomly discarding a full third of the observations, which represents a staggering loss. And these methods are almost as wasteful when used in place of tests based on ranks.

These tests, nevertheless, do have a limited role in the behavioral scientist's statistical tool kit. They are easy to compute, and can be used to get an approximate idea of what a parametric (or rank) test will show with regard to significance. In situations where many tests must be performed and computing facilities are not available, these methods can be used to screen the variables that are highly significant and those that are clearly nonsignificant from the remainder. This procedure is most appropriate when N is quite large, so that the loss of power is tolerable. In almost all cases, however, such "quick and dirty" approximations are solely for the benefit of the researcher's ease in computation. When the results are reported in a scientific journal, the appropriate parametric or rank-based tests should be performed and reported.

We will consider two such methods for approximating statistical significance: the median test for differences in location among two or more independent samples, and the sign test for the difference in location between two matched samples.

The Median Test for Two or More Independent Samples

The null hypothesis to be tested is that k (two or more) populations have identical medians. There must be an independent random sample from each population, but the samples need not be of equal size. The size of sample i is symbolized by N_i, while the total sample size is as usual represented by N. Thus $N = N_1 + N_2 + \cdots + N_k$, where k is equal to the number of groups being compared.

The first step is to determine the median of the total (combined) samples of N cases, which is called the *grand median* (G Mdn). Next count the number of cases in each sample that fall above G Mdn, and the number that fall at or below G Mdn. Finally, organize these frequencies into a $2 \times k$ table and perform a chi-square test of independence, as described in the previous chapter. This test thus assesses the relationship between membership in one of the k groups and "high" versus "not high" on the dependent variable. If the chi-square value is significant, you may conclude that the distribution of the dependent variable is *not* the same across the k populations; one or more of them differ.

---

* Cohen, *op. cit.*

To illustrate, suppose that a national opinion survey of college students is conducted in order to determine the relationship between political affiliation and attitudes toward "Traditional Marriage and the Family" (TMF). A written attitude scale is used to measure TMF, with the scores oriented so that high values indicate a favorable attitude, and each student's political preference is also ascertained. A total sample of $N = 512$ is obtained, of whom $N_1 = 184$ are Democrats, $N_2 = 108$ are Republicans, and $N_3 = 220$ are Independents. The researchers wish to compare the locations of these three groups on TMF, and they decide to use the median test for three reasons: The three distributions of TMF scores are quite positively skewed, making a distribution-free test appropriate; the large sample makes tolerable the loss in power that occurs when this test is used; and they wish a method that is quick and easy to calculate.

The latter stages of the calculation of the median test are shown in Table 18.4. (Although the data used are fictitious, they accord closely with actual pre-Watergate survey findings.) The grand median has already been computed, and the number of cases above G Mdn and the number of cases at or below G Mdn have been determined for each group separately. The reason why the row totals are not identical, even though the sample has been split at the median, is that all cases precisely equal to G Mdn have been placed in the "at or below" category. Table 18.4 reveals that 47.3% (87/184) of the Democrats are "high" on TMF (above G Mdn), as are 66.7% (72/108) of the Republicans, and 40.0% (88/220) of the Independents. Obviously, if you were to determine the median of each group, the three sample medians would differ. The question is whether the three *population* medians differ, that is, whether the three sample medians differ significantly.

For this purpose, the chi-square test for two-variable problems is used exactly as described in Chapter 17. The data in Table 18.4 yield a χ^2 value of 20.74, which is referred to Table G (in the Appendix) with $df = (r - 1)(c - 1)$, where r is the number of rows and c is the number of columns. Therefore, for this problem, $df = (2 - 1)(3 - 1) = 2$. (Since the number of rows in a median test is always equal to 2, the first quantity in

TABLE 18.4

The median test: Relationship of political affiliation to position above or below grand median on TMF attitude scale

TMF Scale	Political Affiliation			
	Democratic	Republican	Independent	Row Total
above G Mdn	87	72	88	247
at or below G Mdn	97	36	132	265
column total (N_i)	184	108	220	$N = 512$
	$\chi^2 = 20.74$,	$df = k - 1 = 2$		

the expression for df is always 1, so the df for the median test can be expressed simply as $c - 1$ or $k - 1$.) This value, 20.74, exceeds the tabled value of 5.99 for the .05 significance criterion, so the null hypothesis of equal population medians should be rejected.

As was the case with chi square, the *strength* of the relationship between the two variables may be determined by computing an index such as C or Cramér's ϕ. For example, using the data in Table 18.4,

$$C = \sqrt{\frac{\chi^2}{N + \chi^2}}$$

$$= \sqrt{\frac{20.74}{512 + 20.74}}$$

$$= .197$$

$$\text{Cramér's } \phi = \sqrt{\frac{\chi^2}{N(k - 1)}}$$

$$= \sqrt{\frac{20.74}{(512)(1)}}$$

$$= .201$$

(Recall that in the above formula, k stands for the *smaller* of the number of rows and number of columns. Thus $k - 1$ is equal to $2 - 1$ or 1, and this will always be the case in a median test.) Using either index, the degree of relationship between political affiliation and the TMF dichotomy is small (although not trivial). Note once again that a large and highly significant statistical value is not necessarily indicative of a strong degree of relationship; it means only that one can confidently assert that the relationship in the population is not zero. The χ^2 value in Table 18.4 was actually significant beyond the .001 level, yet the strength of the relationship between the two variables was only about .20—a not atypical situation when sample sizes are extremely large.

When a statistically significant value of χ^2 is obtained in a median test, multiple comparisons may be performed in order to determine which pairs of populations differ from one another in location. As was the case with ANOVA and H, the requirement that overall significance be obtained affords protection against a large experimentwise Type I error rate. To conduct the multiple comparisons, simply perform a new median test for any or all pairs of groups. That is, for a given pair of groups, find the G Mdn for the combined observations *in those two groups* and perform the χ^2 test on the resulting 2×2 table of frequencies. Note that this new G Mdn will in general be different from the G Mdn of the entire sample of N cases.

Since the overall χ^2 was significant in the illustrative example, median tests may be performed on each of the $k(k - 1)/2 = 3$ pairs of groups. We will illustrate one of these three tests, the comparison between Democrats and Republicans. Their G Mdn on TMF is equal to 28 (larger than the original total-sample G Mdn of 23, since the low-scoring Independents are now excluded), so the frequencies of cases above 28 and at or below 28 are found and organized in a 2 × 2 table:

	D	R	
>28	76	64	140
≤28	108	44	152
N_i	184	108	292 = N

The computation of χ^2 may now proceed, using the formula for 2 × 2 tables given in Chapter 17:

$$\chi^2 = \frac{292(76 \times 44 - 64 \times 108)^2}{(140)(152)(184)(108)}$$

$$= \frac{292(-3568)^2}{422,876,160}$$

$$= 8.791$$

For 2 × 2 tables, $df = 1$, and the obtained value of 8.791 exceeds the critical value of 3.84 necessary for rejecting the null hypothesis at the .05 level of significance. Since 76/184 or 41.3% of the Democrats and 64/108 or 59.3% of the Republicans fell above their G Mdn, you conclude that the population Mdn on TMF for Republicans is higher than that for Democrats.

Since this is a 2 × 2 table, an ordinary phi coefficient may be used as a measure of the strength of the relationship in the sample between D-R and the TMF dichotomy (Chapter 17):

$$\phi = \sqrt{\frac{\chi^2}{N}}$$

$$= \sqrt{\frac{8.791}{292}} = .17$$

This is indicative of a rather weak relationship.

The same analysis applied to the comparison of Republicans and Independents would show a larger χ^2 and ϕ. When Democrats and Independents are compared, the resulting χ^2 is not significant. (The reader cannot check these results, since the three groups' distribution of TMF scores has not been supplied.) The investigator would therefore conclude that Republicans have higher (median) TMF scores than either Democrats or Independents, while the latter two populations do not demonstrably differ.

Since the statistical model employed for the median test is χ^2, the rules about minimum expected frequencies given in Chapter 17 should be observed. This is unlikely to prove a hardship, however, since the statistical power of the median test is so low; using it with samples anywhere near small enough to violate the required minimums would indeed be foolhardy.

The Sign Test for Matched Samples

In the parametric t test for matched samples, the actual values of the paired differences were used. In the Wilcoxon test for matched samples, the information was reduced to the ranks of the values (and their signs). In the sign test for matched samples, however, only the signs of the D values are used. That is, for each of the N pairs, you simply ascertain whether D is positive (as happens when $X_1 > X_2$) or negative (as happens when $X_1 < X_2$), and you count how many pluses and minuses there are. As in the Wilcoxon test, pairs for which $D = 0$ are dropped and N is reduced accordingly. The issue is whether there is a preponderance of either sign, and the formal null hypothesis is that the population Mdn of the D values equals zero (or that positive and negative signs are equiprobable, with $P(+) = P(-) = .50$). Rejection of the null hypothesis leads to the conclusion that there are more positive (negative) Ds, so the location of the first population is higher (lower) than that of the second population.

Since the sign test involves the observed frequencies for positive and negative Ds, and the frequencies that are expected under the null hypothesis ($N/2$ for each), a one-variable χ^2 test may be performed (as in the first part of Chapter 17). The sign test always involves exactly two categories (and thus $df = 1$) and the null hypothesis is that of equiprobability, so a somewhat simpler computational formula for χ^2 is available:

$$\chi^2 = \frac{(f_P - f_m)^2}{N}$$

where

       f_p = number of positive Ds
       f_m = number of negative Ds
       N = total number of pairs (excluding any cases where $D = 0$)

To illustrate, suppose that a consumer research study is conducted to determine preferences between two flavors of a soft drink. The sample consists of 200 subjects, and the .05 criterion of significance is to be used. Each subject tastes each of the two flavors in random order and rates them on a seven-point scale from "awful" to "excellent," and $D = X_1 - X_2$ is computed for each subject. After those who rate the two flavors equally ($D = 0$) are excluded, it is found that the number of positive Ds is 97 while the number of negative Ds is 64. That is, 97 people prefer flavor 1 while 64 people prefer flavor 2, and a total of $97 + 64$ or 161 people have a preference. Then,

$$\chi^2 = \frac{(97 - 64)^2}{161}$$

$$= 6.76$$

When this value is referred to Table G in the Appendix, it is found to exceed the critical value of 3.84 for $df = 1$ and $\alpha = .05$. Therefore the null hypothesis is rejected, and the conclusion is that there is a preponderance of preference for flavor 1 in the population.

Note that the analysis would be identical if, instead of rating each flavor and computing D, each subject were asked simply to designate which flavor was preferred. This would provide all of the information required by the sign test, namely the direction or sign of the difference. When only the direction of the paired differences constitute the basic data of an investigation, there is no alternative to the sign test.

Because chi square is the test statistic (with $df = 1$), small expected frequencies should be avoided, and the sign test should not be used when N is less than 15 or 20. Of course, as we have stressed, the power efficiency of the sign test is such that it should be avoided in any case unless N is quite large.

Summary

When there is reason to believe that the shape of a population distribution is substantially nonnormal, or when a shortcut approximate test is desired, nonparametric or distribution-free statistical tests may be used.

1. Basic considerations

The main advantage of nonparametric and distribution-free statistical tests is that they do *not* require the population(s) being sampled to be normally distributed, so they are applicable when gross nonnormality is suspected. The primary disadvantage of these methods is that when normality does exist, they are less powerful than the corresponding parametric tests (more likely to lead to a Type II error). A numerical measure of the *power efficiency* of a distribution-free or nonparametric test is

$$\frac{N_p}{N_d} \times 100\%$$

where

N_p = sample size required by a parametric test to obtain a specified power for a specified criterion of signficance and a specified difference between population means

N_d = sample size required by the corresponding distribution-free or nonparametric test to obtain the same power under the same conditions

The power efficiency of a nonparametric or distribution-free test is almost always less than 100%, and sometimes much less. Thus it is wasteful to use these methods when parametric tests are applicable.

2. The difference between the locations of two independent samples: The rank-sum test

1. Parametric analog: *t* test for the difference between two independent means (Chapter 11).
   Power efficiency: Approximately 92–95%
2. Computational procedures: Rank *all* scores from 1 (smallest) to N (largest), regardless of which group they are in. In case of ties, follow the usual procedure of assigning the mean of the ranks in question to each of the tied scores. Then compute

$$z = \frac{T_1 - T_E}{\sqrt{\dfrac{N_1 N_2 (N + 1)}{12}}}$$

where

T_1 = sum of ranks in group 1
$T_E = N_1(N + 1)/2$
N_1 = number of observations in group 1
N_2 = number of observations in group 2
N = total number of observations

Do *not* use this procedure if there are fewer than six cases in any group.
3. Measure of strength of relationship: If the value of z computed in the rank-sum test is statistically significant, a measure of the strength of the relationship between group membership and the rank values may be found by computing the *Glass rank biserial correlation* (r_G):

$$r_G = \frac{2(\overline{R}_1 - \overline{R}_2)}{N}$$

where

$$\overline{R}_1 = \text{mean of ranks in group 1}$$
$$\overline{R}_2 = \text{mean of ranks in group 2}$$
$$N = \text{total number of observations}$$

The r_G coefficient is *not* a Pearson r, but does fall between the limits of -1 and $+1$.

3. Differences among the locations of two or more independent samples: The Kruskal-Wallis H test

1. Parametric analog: F test of one-way ANOVA (Chapter 15)
   Power efficiency: Approximately 90–95%
2. Computational procedures: Rank *all* scores from 1 (smallest) to N (largest), regardless of which group they are in. In case of ties, follow the usual procedure of assigning the mean of the ranks in question to each of the tied scores. Then compute the sum of squares between groups (SS_B) for the ranks, using either the usual formula given in Chapter 15 or the following somewhat simpler one that is especially designed for ranked data:

$$SS_B = \frac{T_1^2}{N_1} + \frac{T_2^2}{N_2} + \cdots + \frac{T_k^2}{N_k} - \frac{N(N+1)^2}{4}$$

where

$$T_1 = \text{sum of ranks in group 1}$$
$$T_2 = \text{sum of ranks in group 2}$$
$$T_k = \text{sum of ranks in group } k$$
$$N_1 = \text{number of observations in group 1}$$
$$N_2 = \text{number of observations in group 2}$$
$$N_k = \text{number of observations in group } k$$
$$k = \text{number of groups}$$
$$N = \text{total number of observations}$$

Once SS_B has been obtained, compute

$$H = \frac{12SS_B}{N(N+1)}$$

The value of H is then referred to the χ^2 table with $k - 1$ degrees of freedom. If H equals or exceeds the tabled value, reject the null hypothesis that all populations have equal locations. Otherwise retain H_0. Do *not* use this procedure if there are only two or three groups *and* any group has fewer than five cases.

3. Multiple comparisons: If (and only if) H is statistically significant, the *protected rank-sum test* may be used to determine which pairs of populations differ significantly in location. First the two groups being compared must be *reranked*, with all other groups being ignored for purposes of this comparison. Then, the rank-sum test described previously is computed (with N equal to the number of observations *in these two groups*). This test may be performed for any or all pairs of groups.

4. Measure of strength of relationship: If H is statistically significant, a measure of the strength of relationship between group membership and rank on the dependent variable may be obtained by computing epsilon applied to ranks (ε_R):

$$\varepsilon_R = \sqrt{\frac{H - k + 1}{N - k}}$$

where

k = number of groups
N = total number of observations

4. The difference between the locations of two matched samples: The Wilcoxon test

1. Parametric analog: matched t test (Chapter 11)
   Power efficiency: Approximately 92–95%

2. Computational procedures: First obtain the D score for each subject by subtracting X_2 from X_1. Next, discard any case where $D = 0$ (and reduce N accordingly). Then, *ignoring the sign* of the Ds, rank them from the smallest (rank = 1) to the largest (rank = N). Finally, compute

$$z = \frac{T_1 - T_E}{\sqrt{\dfrac{(2N + 1)T_E}{6}}}$$

where

T_1 = sum of ranks for those D values that are positive (the sum of the ranks for those D values that are negative may instead be used)
$T_E = N(N + 1)/4$
N = number of pairs (excluding cases where $D = 0$)

Do *not* use this procedure if N is less than about 8.

3. Measure of strength of relationship: If the value of z computed in the Wilcoxon test is statistically significant, a measure of the strength of

the relationship between the condition (X_1 versus X_2) and the dependent variable may be found by computing the *matched-pairs rank biserial correlation* (r_C):

$$r_C = \frac{4(T_1 - T_E)}{N(N + 1)}$$

where T_1, T_E, and N have the same meaning as in the Wilcoxon test. The r_C coefficient is *not* a Pearson r, but does fall between the limits of -1 and $+1$.

5. Rough methods for approximating statistical significance: The median and sign tests

The power efficiency of the median and sign tests is only about $65-70\%$ in most cases, so these methods are used primarily to obtain a quick approximation of the results of more powerful (parametric or rank) tests when samples are large.

The median test is used to compare the locations of two or more independent samples. The first step is to compute the grand median (G Mdn) for all N observations. Next organize a $k \times 2$ table, where each sample is divided into two parts: observations falling *above G Mdn,* and observations falling *at or below G Mdn.* Then count and tabulate the frequencies, and compute a two-way chi square (Chapter 17) with $k - 1$ *df.*

The sign test is used to compare the locations of two or more matched samples. First compute D ($= X_1 - X_2$) for each pair. Next, discard all cases where $D = 0$ and reduce N accordingly. Then compute a one-way chi square (Chapter 17), or use the following (simpler) formula that is designed especially for the sign test:

$$\chi^2 = \frac{(f_p - f_m)^2}{N}, \quad df = 1$$

where

$$f_p = \text{number of positive } Ds$$
$$f_m = \text{number of negative } Ds$$
$$N = \text{total number of pairs}$$
$$\text{(excluding cases where } D = 0)$$

Chapter 19
Conclusion

Introduction

We have now reached the end of our journey into the realm of introductory statistics. We hope that you have come to appreciate the importance of this distinctive and challenging area, and that you have found it (perhaps to your pleasant surprise) to be both understandable and interesting.

When you review the material in the preceding pages, do not let the many different procedures cause you to lose sight of the underlying rationale. Keep in mind the major objective of all statistical analyses: The behavioral scientist wants to *study important issues, ask significant questions,* and *unearth useful discoveries* about our complicated and singular species. To attain a better world, we need a better understanding of human behavior, and we believe that the scientific method is the best way to achieve it. Unfortunately, research experiments cannot deal directly with infinite populations, since our resources in time and money are pitifully finite. The scientist must instead work with a relatively small sample of the phenomena to be studied. The question then arises: How can we generalize the results obtained from the sample to the infinite population?

Statistics — in particular, inferential statistics — enables us to answer that question. The statistical method is certainly not infallible. In fact, as we have seen, the methods are probabilistic. Even a well-designed and properly analyzed experiment *may* yield misleading results, as when a Type I error occurs. But certainty is impossible, because measuring all instances of a phenomenon forever is impossible. Statistical procedures allow us to make progress and advance our knowledge, in the face of unavoidable uncertainty.

Common Statistical Errors

One objective of this book has been to help you avoid some all too common errors in statistical analysis — ones that, unfortunately, can be found even in published research articles. Let us therefore highlight the most important of these troublesome pitfalls.

1. Drawing inferences from retaining H_0, especially when N is small. Suppose a researcher interested in personality theory computes the Pearson correlation between a measure of introversion-extroversion and grades in statistics, using a sample of 14 undergraduate students. The result proves to be $r = +.09$, which is not statistically significant. The researcher therefore writes: "Of course H_0 must be retained. We must conclude that introversion-extraversion has little relationship to proficiency in statistics because the correlation is so close to zero."

As you will have noted, the researcher's second sentence represents a gross error. Because of the small sample size, *even if* the correlation in the population were high, the probability of a Type II error is very large — so much so that *no* conclusions, not even a small "weasel," can be drawn from this experiment. The researcher should have written: "Of course H_0 must be retained, since the data do not provide a basis for refuting it. This experiment has shown nothing at all, and no positive conclusions of any kind can be made. In fact, I probably shouldn't have carried it out with so few subjects, since the probability of obtaining statistically significant results *even if the population correlation were* + .5 is low. I'd better reread Chapter 14."

2. Confusing statistical and practical significance. As we have stressed throughout this book, statistical significance does *not* imply that the results have practical utility as well. Instead, it is necessary to convert statistically significant results into measures that express the strength of the relationship between the variables in question, or otherwise get some notion as to how *large* an effect is.

To illustrate, suppose that a diligent researcher obtains a sample of 4,000 young men and 3,500 young women and administers the SAT to all of them. He finds that the mean SAT score for the men is 500.3, the mean for the women is 505.1, and the t test for the difference between two means is 2.07. Since this t value is statistically significant at the .05 level, the researcher writes: "Much to the dismay of my male pride, I must conclude that women are smarter than men." Actually, all that this experiment shows is that the means of two populations are probably never *exactly* equal. The effect of sex on intelligence (as measured by the SAT) is very small, as the researcher would have discovered by converting t to r_{pb} and finding that $r_{pb} = .02$.

Confidence intervals can also help to avoid this error. Suppose the researcher finds that the 95% confidence interval for the SAT experiment is 1.9–7.7. While the difference between the population means is probably not zero, a difference as small as two points is reasonably likely, and such a difference would be negligible for practical purposes. Alternatively, if the 95% confidence interval is 4.1–5.5 (a small interval being likely with such a large sample, which reduces the amount of sampling error), the researcher would be warned that the largest probable difference between the two population means is only $5\frac{1}{2}$ SAT points.

3. Improper generalizations. Even when statistical significance and practical utility are both achieved, the results can safely be generalized only to the population from which the sample was drawn (and randomly drawn, at that). For example, if an experiment is conducted using 50 undergraduate college students enrolled in introductory psychology, the researcher would be on tenuous ground indeed in trying to apply the results to blue-collar industrial workers — or even to college undergraduates *not* enrolled in introductory psychology, or at a college or university in a different geographical location.

4. Collecting unanalyzable data. The statistical design of an experiment must be decided upon *before* the data are collected. This cannot be stressed too strongly! It is highly desirable (1) to state clearly the objectives of the study, (2) to identify and describe the relevant variables, and (3) to specify the sample(s) you will be using and the population to which you wish to generalize. Piloting, or pretesting your procedures on a small sample, allows you to try out and improve your procedures for obtaining subjects, your instructions to them, and the instruments you plan to use, as well as to finalize your plans for the analysis of the data. Otherwise you might discover that because of the form your data are in, or the method by which they were collected, or the omission of relevant data, there are *no* statistical procedures that can be used to analyze them.

The Problem of Experimental Design

Since there is such a large variety of research designs and strategies, it would seem desirable to conclude this book by presenting a list of precise and unambiguous rules for selecting the best ones. Unfortunately, such a discussion would require another entire book. Research design is usually not like a classroom examination problem, where there is one "right answer." Instead, the choice of statistical analysis involves a variety of scientific, practical, and personal considerations. Thus, in the following review, we have attempted only to highlight some of the techniques covered in this textbook.

Assume that you have properly stated your research problem, reviewed the relevant literature, and decided upon your instruments and samples. The examples outlined in Table 19.1 may guide you in deciding how to select an experimental design to test your hypothesis(es). At the same time, we have included a short discussion of potential problems to serve as a guide for you to critically evaluate your design. See if you can find additional potential problems in each research design.

Table 19.1 could have been extended to include correlational methods and nonparametric techniques (including χ^2). Rather than making the table prohibitively long, we have chosen to concentrate on some fundamental design issues. We hope that this discussion has helped to clarify some of these issues.

Now that you have finished this book, you should be able to read the behavioral science literature more critically. We also hope that you will have been encouraged to learn more about statistics in the future.

TABLE 19.1a
Selected research problems

Research Problem	Statement of the Null (H_0) and Alternative (H_1) Hypotheses
You are working in a residential treatment center with $N = 40$ eight-year-old children. You suspect that their vocabulary skills are suffering as a result of their living conditions. You wish to compare these children with eight-year-old children from the general population. You administer a standard vocabulary test to all 40 children and find $\overline{X} = 12$, $s = 3$. Data are available from the scoring manual on the mean (μ) for children of this age in the normative population. This $\mu = 15$.	$H_0: \mu = 15$ $H_1: \mu \neq 15$ H_0 hypothesis is a statement that the difference between the normative population mean (15) and the sample mean (12) is due to sampling error, i.e., these children actually come from a population where $\mu = 15$. The alternate hypothesis is, of course, the negation of the null.

Note: For all the examples, assume that you are using a two-tailed test of the H_0 hypothesis, and that $\alpha = .05$. Assume also that your decision about sample size has been made in the light of a power analysis in which you have posited a reasonable effect size as an alternate to H_0 (see Chapter 14).

TABLE 19.1a

Selected research problems

Statistical Technique and Decision(s)	Potential Problems and Related Issues
You have *one sample and a null hypothesis about the specific value of the population mean.* Use the *t test* for the mean of a single population, σ unknown: $$t = \frac{\overline{X} - \mu}{s_{\overline{X}}} \text{ with } df = N - 1 = 39$$ 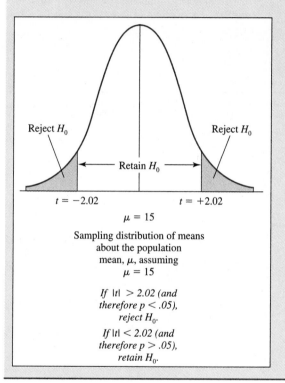 Reject H_0 Reject H_0 ← Retain H_0 → $t = -2.02$ $t = +2.02$ $\mu = 15$ Sampling distribution of means about the population mean, μ, assuming $\mu = 15$ *If |t| > 2.02 (and therefore p < .05), reject H_0.* *If |t| < 2.02 (and therefore p > .05), retain H_0.*	Since you are not dealing with a random sample from the population to which you wish to generalize (i.e., you cannot randomly select children and place them in a residential treatment center), differences may not be due to conditions within the center. Vocabulary may be affected by socioeconomic conditions, medical history, sex of child, etc. Confidence intervals provide important additional information, while also telling you whether to retain or reject the null hypothesis. This technique is rarely used. Population values are seldom available. More typically, you wish to make statements about the differences between two populations tested under different conditions using two samples, one sample from each population (see problem b).

TABLE 19.1b

Selected research problems

Research Problem	Statement of the Null (H_0) and Alternative (H_1) Hypotheses
You are interested in the effects of extra reading assignments on the vocabulary skills of third-grade children attending a public school. Before the new school year you randomly select two groups of children ($N = 25$ per group) from all third-grade children and place one group within a traditional classroom (the control group, C), and the other (the experimental group, E) in a class where they will be given additional reading assignments. A standard vocabulary test is administered to both groups at the end of the term (5 months). You find $\overline{X}_E = 16.8$ and $\overline{X}_C = 15.2$.	$H_0: \mu_E = \mu_C$ $H_1: \mu_E \neq \mu_C$ The null hypothesis states that there is no difference between the vocabulary means of the two populations from which these samples were drawn. (Or, the difference in the means of the two samples is due to sampling error.)

Note: For all the examples, assume that you are using a two-tailed test of the H_0 hypothesis, and that $\alpha = .05$. Assume also that your decision about sample size has been made in the light of a power analysis in which you have posited a reasonable effect size as an alternate to H_0 (see Chapter 14).

TABLE 19.1b

Selected research problems

Statistical Technique and Decision(s)	Potential Problems and Related Issues
You have randomly selected *two independent samples* from the population of third graders at a specific school. Use the <u>t test for differences between two independent samples</u>: $$t = \frac{\bar{X}_1 - \bar{X}_2}{\sqrt{\dfrac{(N_1 - 1)s_1^2 - (N_2 - 1)s_2^2}{N_1 + N_2 - 2}\left(\dfrac{1}{N_1} + \dfrac{1}{N_2}\right)}}$$ with $df = N_1 + N_2 - 2$ $= 25 + 25 - 2$ $= 48$ 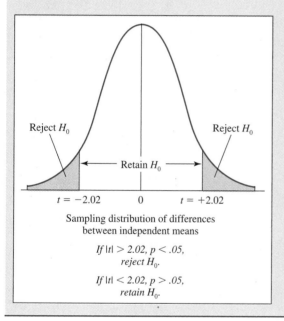 $t = -2.02 \qquad 0 \qquad t = +2.02$ Sampling distribution of differences between independent means *If |t| > 2.02, p < .05, reject H_0.* *If |t| < 2.02, p > .05, retain H_0.*	Are the *same* teachers used for both groups? Is the extra attention given to the experimental group affecting their motivation and therefore their scores? Are the scores of boys and girls affected in the same ways by the different treatment modalities? To what schools do you wish to generalize? (Your samples came from only one public school. The demographic characteristics of children in this school may be quite different from those in other public schools.) If the difference between the means of these two groups is significant, is it meaningful? Explore by transforming the *t* into a point biserial *r* (r_{pb}): $$r_{pb} = \sqrt{\frac{t^2}{t^2 + df}}$$ The closer r_{pb} is to 1.00, the stronger the relationship between group membership and vocabulary score. Confidence intervals provide important additional information, while also telling you whether to retain or reject the null hypothesis.

Inside the figure:

Reject H_0 Reject H_0

Retain H_0

TABLE 19.1c

Selected research problems

Research Problem	Statement of the Null (H_0) and Alternative (H_1) Hypotheses
You are interested in whether teaching children the meaning of specific prefixes and suffixes will affect their vocabulary skills. You randomly select a group of 25 third-grade children and administer a vocabulary test. After a four-week training period where they are taught to identify particular prefixes and suffixes, they are again administered the same test.	$H_0 : \mu_D = 0$ $H_1 : \mu_D \neq 0$

Note: For all the examples, assume that you are using a two-tailed test of the H_0 hypothesis, and that $\alpha = .05$. Assume also that your decision about sample size has been made in the light of a power analysis in which you have posited a reasonable effect size as an alternate to H_0 (see Chapter 14).

Selected research problems

Statistical Technique and Decision(s)	Potential Problems and Related Issues
Since the *same group* is *tested twice,* use the <u>*t* test for matched samples:</u> $$t = \frac{\overline{D} - \mu_D}{\sqrt{\dfrac{s_D^2}{N}}}$$ with *df* = Number of pairs − 1 = 25 − 1 = 24	If the same words are used in testing pre- and post-vocabulary skills, the children may have become familiar with them, reducing the effects of the experimental manipulation. As in problem b, if the difference between the pre- and posttests is significant, is it also meaningful? Again, transform *t* to r_{pb}. As in problems a and b, confidence intervals provide important additional information while also telling you whether to retain or reject the null hypothesis.

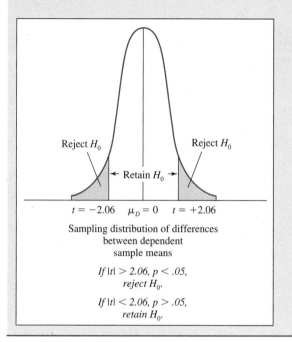

Reject H_0 Reject H_0

← Retain H_0 →

$t = -2.06$ $\mu_D = 0$ $t = +2.06$

Sampling distribution of differences
between dependent
sample means

If |*t*| > 2.06, *p* < .05,
reject H_0.

If |*t*| < 2.06, *p* > .05,
retain H_0.

TABLE 19.1d

Selected research problems

Research Problem	Statement of the Null (H_0) and Alternative (H_1) Hypotheses
You are interested in knowing whether vocabulary scores are affected by different teaching methods. You select 120 third-grade children from all third graders in a public school and randomly place 40 in each of three classroom treatment conditions. Classroom 1 follows traditional rote-learning procedures; in classroom 2 word games are substituted for traditional rote procedures; in classroom 3 children view a series of short educational films focusing on new words. At the end of one month, children are administered a standard vocabulary test.	H_0: $\mu_1 = \mu_2 = \mu_3$ H_1: H_0 taken as a whole is not true H_0 is a statement that there are no differences among the vocabulary scores of the three groups. If differences are not significant, the experimental manipulations have, essentially, not been effective.

Note: For all the examples, assume that you are using a two-tailed test of the H_0 hypothesis, and that $\alpha = .05$. Assume also that your decision about sample size has been made in the light of a power analysis in which you have posited a reasonable effect size as an alternate to H_0 (see Chapter 14).

TABLE 19.1d

Selected research problems

Statistical Technique and Decision(s)	Potential Problems and Related Issues
Since there are *more than two groups,* you must use the <u>Analysis of Variance (One-Way)</u> where: $$F = \frac{MS_{\text{Between Groups}}}{MS_{\text{Within Groups}}} = \frac{MS_B}{MS_W}$$ with $df_B = k - 1$ (where k = the number of groups) $= 3 - 1 = 2$ and $df_W = N - k$ (where N = the total number of observations) $= 120 - 3 = 117$ Look up F with 2 and 117 *df.*	Are the same amounts of time being spent in teaching the children under the three conditions? For example, is more time being spent in word games than in rote learning? Are boys and girls affected in the same way by the experimental conditions? (See problem e.) The overall H_0 is a statement about differences among all three populations. To test differences between any two population means (i.e., 1 vs. 2, 1 vs. 3, 2 vs. 3), use the Protected t test *if F* is significant. A confidence interval may be established for each comparison. If F is significant, is there a strong or weak relationship between group membership and vocabulary? Test strength of relationship using epsilon (ε) where: $$\varepsilon = \sqrt{\frac{df_B\,(F - 1)}{df_B\,F + df_W}}$$

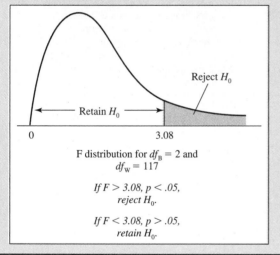

Reject H_0

Retain H_0

0   3.08

F distribution for $df_B = 2$ and
$df_W = 117$

If $F > 3.08$, $p < .05$,
reject H_0.

If $F < 3.08$, $p > .05$,
retain H_0.

Selected research problems

Research Problem	Statement of the Null (H_0) and Alternative (H_1) Hypotheses

The issue is the same as in problem d, but you suspect that boys and girls may react differently under the three conditions. For example, you may feel that audiovisual aids (films) may be more effective for boys than for girls. Therefore, 40 children are again randomly selected for each condition, but within each condition there are 20 boys and 20 girls.	There are a number of hypotheses that can be tested. Two are related to the main effects of (a) experimental condition and (b) sex, and the third involves (c) the interaction of sex and condition:
	
	(a) H_0 (Learning Condition): There are no differences in the population means for the three conditions.
	   H_1: H_0 taken as a whole is untrue.
	(b) H_0 (Sex): There are no differences in the population means of boys and girls.
	   H_1: H_0 is untrue.
	(c) H_0 (Interaction of Sex $\times$ Condition): The differences between the means of vocabulary scores for boys vs. girls in the population are not significantly different across experimental conditions (i.e., differences between the means of the scores of the boys and the means of the scores of the girls are of equal size across conditions).
	   H_1: H_0 is untrue.

Note: For all the examples, assume that you are using a two-tailed test of the H_0 hypothesis, and that $\alpha = .05$. Assume also that your decision about sample size has been made in the light of a power analysis in which you have posited a reasonable effect size as an alternate to H_0 (see Chapter 14).

TABLE 19.1e

Selected research problems

Statistical Technique and Decision(s)	Potential Problems and Related Issues
Since there are *two independent variables or factors* (learning condition and sex,) use the Factorial Design, Two-Way Analysis of Variance. There are three learning conditions *(LC)* and two sexes *(S)* or six groups *(k)* as follows:	If there is an interaction effect, you must modify what you say about the main effects. You may find that the main effect for sex is not significant and conclude (incorrectly if the interaction effect is significant) that there are no differences in vocabulary scores between males and females. Boys may have much higher scores than girls when assigned to the films condition while girls may be clearly superior in the traditional learning condition.

	Conditions		
	Traditional Learning	Word Games	Films
Sex			
Boys	$N = 20$	$N = 20$	$N = 20$
Girls	$N = 20$	$N = 20$	$N = 20$

Then test each of three hypotheses:

$$\text{(a)} \quad F(LC) = \frac{MS_{LC}}{MS_W}$$

with df_{LC} = number of learning conditions − 1
    $= 3 - 1 = 2$
 and $df_W = N - k = 120 - 6 = 114$
Look up F with 2 and 114 df.
Therefore, if $F > 3.08$, $p < .05$, reject H_0 *(LC)*.
        if $F < 3.08$, $p > .05$, retain H_0 *(LC)*.

$$\text{(b)} \quad F(S) = \frac{MS_S}{MS_W}$$

with df_S = number of sexes − 1
    $= 2 - 1 = 1$
and $df_W = N - k = 120 - 6 = 114$
Look up F with 1 and 114 df.
Therefore, if $F > 3.93$, $p < .05$, reject H_0 *(S)*.
        if $F < 3.93$, $p > .05$, retain H_0 *(S)*.

$$\text{(c)} \quad F(S \times LC) = \frac{MS_{S \times LC}}{MS_W}$$

with $df_{S \times LC} = 2 \times 1 = 2$

 and $df_W = N - k = 120 - 6 = 114$
Look up F with 2 and 114 df.
Therefore, if $F > 3.08$, $p < .05$, reject H_0 (interaction).
        if $F < 3.08$, $p > .05$, retain H_0 (interaction).

Strength of association among variables can be tested using epsilon (see problem d).
Additional hypotheses can be tested using the protected t test. If the differences among learning conditions give rise to a significant F (and the interaction is not significant), you may wish to test for possible differences between any two learning conditions. As in problem d, a confidence interval may be established for each comparison.

Appendix

TABLE A

Squares, square roots, and reciprocals of numbers from 1 to 1000

N	N^2	$\sqrt{N}$	$1/N$	N	N^2	$\sqrt{N}$	$1/N$
1	1	1.0000	1.000000	41	1681	6.4031	.024390
2	4	1.4142	.500000	42	1764	6.4807	.023810
3	9	1.7321	.333333	43	1849	6.5574	.023256
4	16	2.0000	.250000	44	1936	6.6332	.022727
5	25	2.2361	.200000	45	2025	6.7082	.022222
6	36	2.4495	.166667	46	2116	6.7823	.021739
7	49	2.6458	.142857	47	2209	6.8557	.021277
8	64	2.8284	.125000	48	2304	6.9282	.020833
9	81	3.0000	.111111	49	2401	7.0000	.020408
10	100	3.1623	.100000	50	2500	7.0711	.020000
11	121	3.3166	.090909	51	2601	7.1414	.019608
12	144	3.4641	.083333	52	2704	7.2111	.019231
13	169	3.6056	.076923	53	2809	7.2801	.018868
14	196	3.7417	.071429	54	2916	7.3485	.018519
15	225	3.8730	.066667	55	3025	7.4162	.018182
16	256	4.0000	.062500	56	3136	7.4833	.017857
17	289	4.1231	.058824	57	3249	7.5498	.017544
18	324	4.2426	.055556	58	3364	7.6158	.017241
19	361	4.3589	.052632	59	3481	7.6811	.016949
20	400	4.4721	.050000	60	3600	7.7460	.016667
21	441	4.5826	.047619	61	3721	7.8102	.016393
22	484	4.6904	.045455	62	3844	7.8740	.016129
23	529	4.7958	.043478	63	3969	7.9373	.015873
24	576	4.8990	.041667	64	4096	8.0000	.015625
25	625	5.0000	.040000	65	4225	8.0623	.015385
26	676	5.0990	.038462	66	4356	8.1240	.015152
27	729	5.1962	.037037	67	4489	8.1854	.014925
28	784	5.2915	.035714	68	4624	8.2462	.014706
29	841	5.3852	.034483	69	4761	8.3066	.014493
30	900	5.4772	.033333	70	4900	8.3666	.014286
31	961	5.5678	.032258	71	5041	8.4261	.014085
32	1024	5.6569	.031250	72	5184	8.4853	.013889
33	1089	5.7446	.030303	73	5329	8.5440	.013699
34	1156	5.8310	.029412	74	5476	8.6023	.013514
35	1225	5.9161	.028571	75	5625	8.6603	.013333
36	1296	6.0000	.027778	76	5776	8.7178	.013158
37	1369	6.0828	.027027	77	5929	8.7750	.012987
38	1444	6.1644	.026316	78	6084	8.8318	.012821
39	1521	6.2450	.025641	79	6241	8.8882	.012658
40	1600	6.3246	.025000	80	6400	8.9443	.012500

TABLE A
(Continued)

N	N²	√N	1/N	N	N²	√N	1/N
81	6561	9.0000	.012346	121	14641	11.0000	.00826446
82	6724	9.0554	.012195	122	14884	11.0454	.00819672
83	6889	9.1104	.012048	123	15129	11.0905	.00813008
84	7056	9.1652	.011905	124	15376	11.1355	.00806452
85	7225	9.2195	.011765	125	15625	11.1803	.00800000
86	7396	9.2736	.011628	126	15876	11.2250	.00793651
87	7569	9.3274	.011494	127	16129	11.2694	.00787402
88	7744	9.3808	.011364	128	16384	11.3137	.00781250
89	7921	9.4340	.011236	129	16641	11.3578	.00775194
90	8100	9.4868	.011111	130	16900	11.4018	.00769231
91	8281	9.5394	.010989	131	17161	11.4455	.00763359
92	8464	9.5917	.010870	132	17424	11.4891	.00757576
93	8649	9.6437	.010753	133	17689	11.5326	.00751880
94	8836	9.6954	.010638	134	17956	11.5758	.00746269
95	9025	9.7468	.010526	135	18225	11.6190	.00740741
96	9216	9.7980	.010417	136	18496	11.6619	.00735294
97	9409	9.8489	.010309	137	18769	11.7047	.00729927
98	9604	9.8995	.010204	138	19044	11.7473	.00724638
99	9801	9.9499	.010101	139	19321	11.7898	.00719424
100	10000	10.0000	.010000	140	19600	11.8322	.00714286
101	10201	10.0499	.00990099	141	19881	11.8743	.00709220
102	10404	10.0995	.00980392	142	20164	11.9164	.00704225
103	10609	10.1489	.00970874	143	20449	11.9583	.00699301
104	10816	10.1980	.00961538	144	20736	12.0000	.00694444
105	11025	10.2470	.00952381	145	21025	12.0416	.00689655
106	11236	10.2956	.00943396	146	21316	12.0830	.00684932
107	11449	10.3441	.00934579	147	21609	12.1244	.00680272
108	11664	10.3923	.00925926	148	21904	12.1655	.00675676
109	11881	10.4403	.00917431	149	22201	12.2066	.00671141
110	12100	10.4881	.00909091	150	22500	12.2474	.00666667
111	12321	10.5357	.00900901	151	22801	12.2882	.00662252
112	12544	10.5830	.00892857	152	23104	12.3288	.00657895
113	12769	10.6301	.00884956	153	23409	12.3693	.00653595
114	12996	10.6771	.00877193	154	23716	12.4097	.00649351
115	13225	10.7238	.00869565	155	24025	12.4499	.00645161
116	13456	10.7703	.00862069	156	24336	12.4900	.00641026
117	13689	10.8167	.00854701	157	24649	12.5300	.00636943
118	13924	10.8628	.00847458	158	24964	12.5698	.00632911
119	14161	10.9087	.00840336	159	25281	12.6095	.00628931
120	14400	10.9545	.00833333	160	25600	12.6491	.00625000

TABLE A

(Continued)

N	N²	√N	1/N	N	N²	√N	1/N
161	25921	12.6886	.00621118	201	40401	14.1774	.00497512
162	26244	12.7279	.00617284	202	40804	14.2127	.00495050
163	26569	12.7671	.00613497	203	41209	14.2478	.00492611
164	26896	12.8062	.00609756	204	41616	14.2829	.00490196
165	27225	12.8452	.00606061	205	42025	14.3178	.00487805
166	27556	12.8841	.00602410	206	42436	14.3527	.00485437
167	27889	12.9228	.00598802	207	42849	14.3875	.00483092
168	28224	12.9615	.00595238	208	43264	14.4222	.00480769
169	28561	13.0000	.00591716	209	43681	14.4568	.00478469
170	28900	13.0384	.00588235	210	44100	14.4914	.00476190
171	29241	13.0767	.00584795	211	44521	14.5258	.00473934
172	29584	13.1149	.00581395	212	44944	14.5602	.00471698
173	29929	13.1529	.00578035	213	45369	14.5945	.00469484
174	30276	13.1909	.00574713	214	45796	14.6287	.00467290
175	30625	13.2288	.00571429	215	46225	14.6629	.00465116
176	30976	13.2665	.00568182	216	46656	14.6969	.00462963
177	31329	13.3041	.00564972	217	47089	14.7309	.00460829
178	31684	13.3417	.00561798	218	47524	14.7648	.00458716
179	32041	13.3791	.00558659	219	47961	14.7986	.00456621
180	32400	13.4164	.00555556	220	48400	14.8324	.00454545
181	32761	13.4536	.00552486	221	48841	14.8661	.00452489
182	33124	13.4907	.00549451	222	49284	14.8997	.00450450
183	33489	13.5277	.00546448	223	49729	14.9332	.00448430
184	33856	13.5647	.00543478	224	50176	14.9666	.00446429
185	34225	13.6015	.00540541	225	50625	15.0000	.00444444
186	34596	13.6382	.00537634	226	51076	15.0333	.00442478
187	34969	13.6748	.00534759	227	51529	15.0665	.00440529
188	35344	13.7113	.00531915	228	51984	15.0997	.00438596
189	35721	13.7477	.00529101	229	52441	15.1327	.00436681
190	36100	13.7840	.00526316	230	52900	15.1658	.00434783
191	36481	13.8203	.00523560	231	53361	15.1987	.00432900
192	36864	13.8564	.00520833	232	53824	15.2315	.00431034
193	37249	13.8924	.00518135	233	54289	15.2643	.00429185
194	37636	13.9284	.00515464	234	54756	15.2971	.00427350
195	38025	13.9642	.00512821	235	55225	15.3297	.00425532
196	38416	14.0000	.00510204	236	55696	15.3623	.00423729
197	38809	14.0357	.00507614	237	56169	15.3948	.00421941
198	39204	14.0712	.00505051	238	56644	15.4272	.00420168
199	39601	14.1067	.00502513	239	57121	15.4596	.00418410
200	40000	14.1421	.00500000	240	57600	15.4919	.00416667

TABLE A
(Continued)

N	N²	√N	1/N	N	N²	√N	1/N
241	58081	15.5242	.00414938	281	78961	16.7631	.00355872
242	58564	15.5563	.00413223	282	79524	16.7929	.00354610
243	59049	15.5885	.00411523	283	80089	16.8226	.00353357
244	59536	15.6205	.00409836	284	80656	16.8523	.00352113
245	60025	15.6525	.00408163	285	81225	16.8819	.00350877
246	60516	15.6844	.00406504	286	81796	16.9115	.00349650
247	61009	15.7162	.00404858	287	82369	16.9411	.00348432
248	61504	15.7480	.00403226	288	82944	16.9706	.00347222
249	62001	15.7797	.00401606	289	83521	17.0000	.00346021
250	62500	15.8114	.00400000	290	84100	17.0294	.00344828
251	63001	15.8430	.00398406	291	84681	17.0587	.00343643
252	63504	15.8745	.00396825	292	86264	17.0880	.00342466
253	64009	15.9060	.00395257	293	85849	17.1172	.00341297
254	64516	15.9371	.00393701	294	86436	17.1464	.00340136
255	65025	15.9687	.00392157	295	87025	17.1756	.00338983
256	65536	16.0000	.00390625	296	87616	17.2047	.00337838
257	66049	16.0312	.00389105	297	88209	17.2337	.00336700
258	66564	16.0624	.00387597	298	88804	17.2627	.00335570
259	67081	16.0935	.00386100	299	89401	17.2916	.00334448
260	67600	16.1245	.00384615	300	90000	17.3205	.00333333
261	68121	16.1555	.00383142	301	90601	17.3494	.00332226
262	68644	16.1864	.00381679	302	91204	17.3781	.00331126
263	69169	16.2173	.00380228	303	91809	17.4069	.00330033
264	69696	16.2481	.00378788	304	92416	17.4356	.00328947
265	70225	16.2788	.00377358	305	93025	17.4642	.00327869
266	70756	16.3095	.00375940	306	93636	17.4929	.00326797
267	71289	16.3401	.00374532	307	94249	17.5214	.00325733
268	71824	16.3707	.00373134	308	94864	17.5499	.00324675
269	72361	16.4012	.00371747	309	95481	17.5784	.00323625
270	72900	16.4317	.00370370	310	96100	17.6068	.00322581
271	73441	16.4621	.00369004	311	96721	17.6352	.00321543
272	73984	16.4924	.00367647	312	97344	17.6635	.00320513
273	74529	16.5227	.00366300	313	97969	17.6918	.00319489
274	75076	16.5529	.00364964	314	98596	17.7200	.00318471
275	75625	16.5831	.00363636	315	99225	17.7482	.00317460
276	76176	16.6132	.00362319	316	99856	17.7764	.00316456
277	76729	16.6433	.00361011	317	100489	17.8045	.00315457
278	77284	16.6733	.00359712	318	101124	17.8326	.00314465
279	77841	16.7033	.00358423	319	101761	17.8606	.00313480
280	78400	16.7332	.00357143	320	102400	17.8885	.00312500

TABLE A
(Continued)

N	N^2	$\sqrt{N}$	$1/N$	N	N^2	$\sqrt{N}$	$1/N$
321	103041	17.9165	.00311526	361	130321	19.0000	.00277008
322	103684	17.9444	.00310559	362	131044	19.0263	.00276243
323	104329	17.9722	.00309598	363	131769	19.0526	.00275482
324	104976	18.0000	.00308642	364	132496	19.0788	.00274725
325	105625	18.0278	.00307692	365	133225	19.1050	.00273973
326	106276	18.0555	.00306748	366	133956	19.1311	.00273224
327	106929	18.0831	.00305810	367	134689	19.1572	.00272480
328	107584	18.1108	.00304878	368	135424	19.1833	.00271739
329	108241	18.1384	.00303951	369	136161	19.2094	.00271003
330	108900	18.1659	.00303030	370	136900	19.2354	.00270270
331	109561	18.1934	.00302115	371	137641	19.2614	.00269542
332	110224	18.2209	.00301205	372	138384	19.2873	.00268817
333	110889	18.2483	.00300300	373	139129	19.3132	.00268097
334	111556	18.2757	.00299401	374	139876	19.3391	.00267380
335	112225	18.3030	.00298507	375	140625	19.3649	.00266667
336	112896	18.3303	.00297619	376	141376	19.3907	.00265957
337	113569	18.3576	.00296736	377	142129	19.4165	.00265252
338	114244	18.3848	.00295858	378	142884	19.4422	.00264550
339	114921	18.4120	.00294985	379	143641	19.4679	.00263852
340	115600	18.4391	.00294118	380	144400	19.4936	.00263158
341	116281	18.4662	.00293255	381	145161	19.5192	.00262467
342	116964	18.4932	.00292398	382	145924	19.5448	.00261780
343	117649	18.5203	.00291545	383	146689	19.5704	.00261097
344	118336	18.5472	.00290698	384	147456	19.5959	.00260417
345	119025	18.5742	.00289855	385	148225	19.6214	.00259740
346	119716	18.6011	.00289017	386	148996	19.6469	.00259067
347	120409	18.6279	.00288184	387	149769	19.6723	.00258398
348	121104	18.6548	.00287356	388	150544	19.6977	.00257732
349	121801	18.6815	.00286533	389	151321	19.7231	.00257069
350	122500	18.7083	.00285714	390	152100	19.7484	.00256410
351	123201	18.7350	.00284900	391	152881	19.7737	.00255754
352	123904	18.7617	.00284091	392	153664	19.7990	.00255102
353	124609	18.7883	.00283286	393	154449	19.8242	.00254453
354	125316	18.8149	.00282486	394	155236	19.8494	.00253807
355	126025	18.8414	.00281690	395	156025	19.8746	.00253165
356	126736	18.8680	.00280899	396	156816	19.8997	.00252525
357	127449	18.8944	.00280112	397	157609	19.9249	.00251889
358	128164	18.9209	.00279330	398	158404	19.9499	.00251256
359	128881	18.9473	.00278552	399	159201	19.9750	.00250627
360	129600	18.9737	.00277778	400	160000	20.0000	.00250000

TABLE A

(Continued)

N	N^2	$\sqrt{N}$	$1/N$	N	N^2	$\sqrt{N}$	$1/N$
401	160801	20.0250	.00249377	441	194481	21.0000	.00226757
402	161604	20.0499	.00248756	442	195364	21.0238	.00226244
403	162409	20.0749	.00248139	443	196249	21.0476	.00225734
404	163216	20.0998	.00247525	444	197136	21.0713	.00225225
405	164025	20.1246	.00246914	445	198025	21.0950	.00224719
406	164836	20.1494	.00246305	446	198916	21.1187	.00224215
407	165649	20.1742	.00245700	447	199809	21.1424	.00223714
408	166464	20.1990	.00245098	448	200704	21.1660	.00223214
409	167281	20.2237	.00244499	449	201601	21.1896	.00222717
410	168100	20.2485	.00243902	450	202500	21.2132	.00222222
411	168921	20.2731	.00243309	451	203401	21.2368	.00221729
412	169744	20.2978	.00242718	452	204304	21.2603	.00221239
413	170569	20.3224	.00242131	453	205209	21.2838	.00220751
414	171396	20.3470	.00241546	454	206116	21.3073	.00220264
415	172225	20.3715	.00240964	455	207025	21.3307	.00219780
416	173056	20.3961	.00240385	456	207936	21.3542	.00219298
417	173889	20.4206	.00239808	457	208849	21.3776	.00218818
418	174724	20.4450	.00239234	458	209764	21.4009	.00218341
419	175561	20.4695	.00238663	459	210681	21.4243	.00217865
420	176400	20.4939	.00238095	460	211600	21.4476	.00217391
421	177241	20.5183	.00237530	461	212521	21.4709	.00216920
422	178084	20.5426	.00236967	462	213444	21.4942	.00216450
423	178929	20.5670	.00236407	463	214369	21.5174	.00215983
424	179776	20.5913	.00235849	464	215296	21.5407	.00215517
425	180625	20.6155	.00235294	465	216225	21.5639	.00215054
426	181476	20.6398	.00234742	466	217156	21.5870	.00214592
427	182329	20.6640	.00234192	467	218089	21.6102	.00214133
428	183184	20.6882	.00233645	468	219024	21.6333	.00213675
429	184041	20.7123	.00233100	469	219961	21.6564	.00213220
430	184900	20.7364	.00232558	470	220900	21.6795	.00212766
431	185761	20.7605	.00232019	471	221841	21.7025	.00212314
432	186624	20.7846	.00231481	472	222784	21.7256	.00211864
433	187489	20.8087	.00230947	473	223729	21.7486	.00211416
434	188356	20.8327	.00230415	474	224676	21.7715	.00210970
435	189225	20.8567	.00229885	475	225625	21.7945	.00210526
436	190096	20.8806	.00229358	476	226576	21.8174	.00210084
437	190969	20.9045	.00228833	477	227529	21.8403	.00209644
438	191844	20.9284	.00228311	478	228484	21.8632	.00209205
439	192721	20.9523	.00227790	479	229441	21.8861	.00208768
440	193600	20.9762	.00227273	480	230400	21.9089	.00208333

TABLE A
(Continued)

N	N^2	$\sqrt{N}$	$1/N$	N	N^2	$\sqrt{N}$	$1/N$
481	231361	21.9317	.00207900	521	271441	22.8254	.00191939
482	232324	21.9545	.00207469	522	272484	22.8473	.00191571
483	233289	21.9773	.00207039	523	273529	22.8692	.00191205
484	234256	22.0000	.00206612	524	274576	22.8910	.00190840
485	235225	22.0227	.00206186	525	275625	22.9129	.00190476
486	236196	22.0454	.00205761	526	276676	22.9347	.00190114
487	237169	22.0681	.00205339	527	277729	22.9565	.00189753
488	238144	22.0907	.00204918	528	278784	22.9783	.00189394
489	239121	22.1133	.00204499	529	279841	23.0000	.00189036
490	240100	22.1359	.00204082	530	280900	23.0217	.00188679
491	241081	22.1585	.00203666	531	281961	23.0434	.00188324
492	242064	22.1811	.00203252	532	283024	23.0651	.00187970
493	243049	22.2036	.00202840	533	284089	23.0868	.00187617
494	244036	22.2261	.00202429	534	285156	23.1084	.00187266
495	245025	22.2486	.00202020	535	286225	23.1301	.00186916
496	246016	22.2711	.00201613	536	287296	23.1517	.00186567
497	247009	22.2935	.00201207	537	288369	23.1733	.00186220
498	248004	22.3159	.00200803	538	289444	23.1948	.00185874
499	249001	22.3383	.00200401	539	290521	23.2164	.00185529
500	250000	22.3607	.00200000	540	291600	23.2379	.00185185
501	251001	22.3830	.00199601	541	292681	23.2594	.00184843
502	252004	22.4054	.00199203	542	293764	23.2809	.00184502
503	253009	22.4277	.00198807	543	294849	23.3024	.00184162
504	254016	22.4499	.00198413	544	295936	23.3238	.00183824
505	255025	22.4722	.00198020	545	297025	23.3452	.00183486
506	256036	22.4944	.00197628	546	298116	23.3666	.00183150
507	257049	22.5167	.00197239	547	299209	23.3880	.00182815
508	258064	22.5389	.00196850	548	300304	23.4094	.00182482
509	259081	22.5610	.00196464	549	301401	23.4307	.00182149
510	260100	22.5832	.00196078	550	302500	23.4521	.00181818
511	261121	22.6053	.00195695	551	303601	23.4734	.00181488
512	262144	22.6274	.00195312	552	304704	23.4947	.00181159
513	263169	22.6495	.00194932	553	305809	23.5160	.00180832
514	264196	22.6716	.00194553	554	306916	23.5372	.00180505
515	265225	22.6936	.00194175	555	308025	23.5584	.00180180
516	266256	22.7156	.00193798	556	309136	23.5797	.00179856
517	267289	22.7376	.00193424	557	310249	23.6008	.00179533
518	268324	22.7596	.00193050	558	311364	23.6220	.00179211
519	269361	22.7816	.00192678	559	312481	23.6432	.00178891
520	270400	22.8035	.00192308	560	313600	23.6643	.00178571

TABLE A

(Continued)

N	N²	√N	1/N	N	N²	√N	1/N
561	314721	23.6854	.00178253	601	361201	24.5153	.00166389
562	315844	23.7065	.00177936	602	362404	24.5357	.00166113
563	316969	23.7276	.00177620	603	363609	24.5561	.00165837
564	318096	23.7487	.00177305	604	364816	24.5764	.00165563
565	319225	23.7697	.00176991	605	366025	24.5967	.00165289
566	320356	23.7908	.00176678	606	367236	24.6171	.00165017
567	321489	23.8118	.00176367	607	368449	24.6374	.00164745
568	322624	23.8328	.00176056	608	369664	24.6577	.00164474
569	323761	23.8537	.00175747	609	370881	24.6779	.00164204
570	324900	23.8747	.00175439	610	372100	24.6982	.00163934
571	326041	23.8956	.00175131	611	373321	24.7184	.00163666
572	327184	23.9165	.00174825	612	374544	24.7386	.00163399
573	328329	23.9374	.00174520	613	375769	24.7588	.00163132
574	329476	23.9583	.00174216	614	376996	24.7790	.00162866
575	330625	23.9792	.00173913	615	378225	24.7992	.00162602
576	331776	24.0000	.00173611	616	379456	24.8193	.00162338
577	332929	24.0208	.00173310	617	380689	24.8395	.00162075
578	334084	24.0416	.00173010	618	381924	24.8596	.00161812
579	335241	24.0624	.00172712	619	383161	24.8797	.00161551
580	336400	24.0832	.00172414	620	384400	24.8998	.00161290
581	337561	24.1039	.00172117	621	385641	24.9199	.00161031
582	338724	24.1247	.00171821	622	386884	24.9399	.00160772
583	339889	24.1454	.00171527	623	388129	24.9600	.00160514
584	341056	24.1661	.00171233	624	389376	24.9800	.00160256
585	342225	24.1868	.00170940	625	390625	25.0000	.00160000
586	343396	24.2074	.00170648	626	391876	25.0200	.00159744
587	344569	24.2281	.00170358	627	393129	25.0400	.00159490
588	345744	24.2487	.00170068	628	394384	25.0599	.00159236
589	346921	24.2693	.00169779	629	395641	25.0799	.00158983
590	348100	24.2899	.00169492	630	396900	25.0998	.00158730
591	349281	24.3105	.00169205	631	398161	25.1197	.00158479
592	350464	24.3311	.00168919	632	399424	25.1396	.00158228
593	351649	24.3516	.00168634	633	400689	25.1595	.00157978
594	352836	24.3721	.00168350	634	401956	25.1794	.00157729
595	354025	24.3926	.00168067	635	403225	25.1992	.00157480
596	355216	24.4131	.00167785	636	404496	25.2190	.00157233
597	356409	24.4336	.00167504	637	405769	25.2389	.00156986
598	357604	24.4540	.00167224	638	407044	25.2587	.00156740
599	358801	24.4745	.00166945	639	408321	25.2784	.00156495
600	360000	24.4949	.00166667	640	409600	25.2982	.00156250

TABLE A

(Continued)

N	N²	√N	1/N	N	N²	√N	1/N
641	410881	25.3180	.00156006	681	463761	26.0960	.00146843
642	412164	25.3377	.00155763	682	465124	26.1151	.00146628
643	413449	25.3574	.00155521	683	466489	26.1343	.00146413
644	414736	25.3772	.00155280	684	467856	26.1534	.00146199
645	416025	25.3969	.00155039	685	469225	26.1725	.00145985
646	417316	25.4165	.00154799	686	470596	26.1916	.00145773
647	418609	25.4362	.00154560	687	471969	26.2107	.00145560
648	419904	25.4558	.00154321	688	473344	26.2298	.00145349
649	421201	25.4755	.00154083	689	474721	262488	.00145138
650	422500	25.4951	.00153846	690	476100	26.2679	.00144928
651	423801	25.5147	.00153610	691	477481	26.2869	.00144718
652	425104	25.5343	.00153374	692	478864	26.3059	.00144509
653	426409	25.5539	.00153139	693	480249	26.3249	.00144300
654	427716	25.5734	.00152905	694	481636	26.3439	.00144092
655	429025	25.5930	.00152672	695	483025	26.3629	.00143885
656	430336	25.6125	.00152439	696	484416	26.3818	.00143678
657	431649	25.6320	.00152207	697	485809	26.4008	.00143472
658	432964	25.6515	.00151976	698	487204	26.4197	.00143266
659	434281	25.6710	.00151745	699	488601	26.4386	.00143062
660	435600	25.6905	.00151515	700	490000	26.4575	.00142857
661	436921	25.7099	.00151286	701	491401	26.4764	.00142653
662	438244	25.7294	.00151057	702	492804	26.4953	.00142450
663	439569	25.7488	.00150830	703	494209	26.5141	.00142248
664	440896	25.7682	.00150602	704	495616	26.5330	.00142045
665	442225	25.7876	.00150376	705	497025	26.5518	.00141844
666	443556	25.8070	.00150150	706	498436	26.5707	.00141643
667	444889	25.8263	.00149925	707	499849	26.5895	.00141443
668	446224	25.8457	.00149701	708	501264	26.6083	.00141243
669	447561	25.8650	.00149477	709	502681	26.6271	.00141044
670	448900	25.8844	.00149254	710	504100	26.6458	.00140845
671	450241	25.9037	.00149031	711	505521	26.6646	.00140647
672	451584	25.9230	.00148810	712	506944	26.6833	.00140449
673	452929	25.9422	.00148588	713	508369	26.7021	.00140252
674	454276	25.9615	.00148368	714	509796	26.7208	.00140056
675	455625	25.9808	.00148148	715	511225	26.7395	.00139860
676	456976	26.0000	.00147929	716	512656	26.7582	.00139665
677	458329	26.0192	.00147710	717	514089	26.7769	.00139470
678	459684	26.0384	.00147493	718	515524	26.7955	.00139276
679	461041	26.0576	.00147275	719	516961	26.8142	.00139082
680	462400	26.0768	.00147059	720	518400	26.8328	.00138889

TABLE A

(Continued)

N	N^2	$\sqrt{N}$	$1/N$	N	N^2	$\sqrt{N}$	$1/N$
721	519841	26.8514	.00138696	761	579121	27.5862	.00131406
722	521284	26.8701	.00138504	762	580644	27.6043	.00131234
723	522729	26.8887	.00138313	763	582169	27.6225	.00131062
724	524176	26.9072	.00138122	764	583696	27.6405	.00130890
725	525625	26.9258	.00137931	765	585225	27.6586	.00130719
726	527076	26.9444	.00137741	766	586756	27.6767	.00130548
727	528529	26.9629	.00137552	767	588289	27.6948	.00130378
728	529984	26.9815	.00137363	768	589824	27.7128	.00130208
729	531441	27.0000	.00137174	769	591361	27.7308	.00130039
730	532900	27.0185	.00136986	770	592900	27.7489	.00129870
731	534361	27.0370	.00136799	771	594441	27.7669	.00129702
732	535824	27.0555	.00136612	772	595984	27.7849	.00129534
733	537289	27.0740	.00136426	773	597529	27.8029	.00129366
734	538756	27.0924	.00136240	774	599076	27.8209	.00129199
735	540225	27.1109	.00136054	775	600625	27.8388	.00129032
736	541696	27.1293	.00135870	776	602176	27.8568	.00128866
737	543169	27.1477	.00135685	777	603729	27.8747	.00128700
738	544644	27.1662	.00135501	778	605284	27.8927	.00128535
739	546121	27.1846	.00135318	779	606841	27.9106	.00128370
740	547600	27.2029	.00135135	780	608400	27.9285	.00128205
741	549081	27.2213	.00134953	781	609961	27.9464	.00128041
742	550564	27.2397	.00134771	782	611524	27.9643	.00127877
743	552049	27.2580	.00134590	783	613089	27.9821	.00127714
744	553536	27.2764	.00134409	784	614656	28.0000	.00127551
745	555025	27.2947	.00134228	785	616225	28.0179	.00127389
746	556516	27.3130	.00134048	786	617796	28.0357	.00127226
747	558009	27.3313	.00133869	787	619369	28.0535	.00127065
748	559504	27.3496	.00133690	788	620944	28.0713	.00126904
749	561001	27.3679	.00133511	789	622521	28.0891	.00126743
750	562500	27.3861	.00133333	790	624100	28.1069	.00126582
751	564001	27.4044	.00133156	791	625681	28.1247	.00126422
752	565504	27.4226	.00132979	792	627264	28.1425	.00126263
753	567009	27.4408	.00132802	793	628849	28.1603	.00126103
754	568516	27.4591	.00132626	794	630436	28.1780	.00125945
755	570025	27.4773	.00132450	795	632025	28.1957	.00125786
756	571536	27.4955	.00132275	796	633616	28.2135	.00125628
757	573049	27.5136	.00132100	797	635209	28.2312	.00125471
758	574564	27.5318	.00131926	798	636804	28.2489	.00125313
759	576081	27.5500	.00131752	799	638401	28.2666	.00125156
760	577600	27.5681	.00131579	800	640000	28.2843	.00125000

TABLE A

(Continued)

N	N²	√N	1/N	N	N²	√N	1/N
801	641601	28.3019	.00124844	841	707281	29.0000	.00118906
802	643204	28.3196	.00124688	842	708964	29.0172	.00118765
803	644809	28.3373	.00124533	843	710649	29.0345	.00118624
804	646416	28.3549	.00124378	844	712336	29.0517	.00118483
805	648025	28.3725	.00124224	845	714025	29.0689	.00118343
806	649636	28.3901	.00124069	846	715716	29.0861	.00118203
807	651249	28.4077	.00123916	847	717409	29.1033	.00118064
808	652864	28.4253	.00123762	848	719104	29.1204	.00117925
809	654481	28.4429	.00123609	849	720801	29.1376	.00117786
810	656100	28.4605	.00123457	850	722500	29.1548	.00117647
811	657721	28.4781	.00123305	851	724201	29.1719	.00117509
812	659344	28.4956	.00123153	852	725904	29.1890	.00117371
813	660969	28.5132	.00123001	853	727609	29.2062	.00117233
814	662596	28.5307	.00122850	854	729316	29.2233	.00117096
815	664225	28.5482	.00122699	855	731025	29.2404	.00116959
816	665856	28.5657	.00122549	856	732736	29.2575	.00116822
817	667489	28.5832	.00122399	857	734449	29.2746	.00116686
818	669124	28.6007	.00122249	858	736164	29.2916	.00116550
819	670761	28.6182	.00122100	859	737881	29.3087	.00116414
820	672400	28.6356	.00121951	860	739600	29.3258	.00116279
821	674041	28.6531	.00121803	861	741321	29.3428	.00116144
822	675684	28.6705	.00121655	862	743044	29.3598	.00116009
823	677329	28.6880	.00121507	863	744769	29.3769	.00115875
824	678976	28.7054	.00121359	864	746496	29.3939	.00115741
825	680625	28.7228	.00121212	865	748225	29.4109	.00115607
826	682276	28.7402	.00121065	866	749956	29.4279	.00115473
827	683929	28.7576	.00120919	867	751689	29.4449	.00115340
828	685584	28.7750	.00120773	868	753424	29.4618	.00115207
829	687241	28.7924	.00120627	869	755161	29.4788	.00115075
830	688900	28.8097	.00120482	870	756900	29.4958	.00114943
831	690561	28.8271	.00120337	871	758641	29.5127	.00114811
832	692224	28.8444	.00120192	872	760384	29.5296	.00114679
833	693889	28.8617	.00120048	873	762129	29.5466	.00114548
834	695556	28.8791	.00119904	874	763876	29.5635	.00114416
835	697225	28.8964	.00119760	875	765625	29.5804	.00114286
836	698896	28.9137	.00119617	876	767376	29.5973	.00114155
837	700569	28.9310	.00119474	877	769129	29.6142	.00114025
838	702244	28.9482	.00119332	878	770884	29.6311	.00113895
839	703921	28.9655	.00119190	879	772641	29.6479	.00113766
840	705600	28.9828	.00119048	880	774400	29.6648	.00113636

TABLE A

(Continued)

N	N^2	$\sqrt{N}$	$1/N$	N	N^2	$\sqrt{N}$	$1/N$
881	776161	29.6816	.00113507	921	848241	30.3480	.00108578
882	777924	29.6985	.00113379	922	850084	30.3645	.00108460
883	779689	29.7153	.00113250	923	851929	30.3809	.00108342
884	781456	29.7321	.00113122	924	853776	30.3974	.00108225
885	783225	29.7489	.00112994	925	855625	30.4138	.00108108
886	784996	29.7658	.00112867	926	857476	30.4302	.00107991
887	786769	29.7825	.00112740	927	859329	30.4467	.00107875
888	788544	29.7993	.00112613	928	861184	30.4631	.00107759
889	790321	29.8161	.00112486	929	863041	30.4795	.00107643
890	792100	29.8329	.00112360	930	864900	30.4959	.00107527
891	793881	29.8496	.00112233	931	866761	30.5123	.00107411
892	795664	29.8664	.00112108	932	868624	30.5287	.00107296
893	797449	29.8831	.00111982	933	870489	30.5450	.00107181
894	799236	29.8998	.00111857	934	872356	30.5614	.00107066
895	801025	29.9166	.00111732	935	874225	30.5778	.00106952
896	802816	29.9333	.00111607	936	876096	30.5941	.00106838
897	804609	29.9500	.00111483	937	877969	30.6105	.00106724
898	806404	29.9666	.00111359	938	879844	30.6268	.00106610
899	808201	29.9833	.00111235	939	881721	30.6431	.00106496
900	810000	30.0000	.00111111	940	883600	30.6594	.00106383
901	811801	30.0167	.00110988	941	885481	30.6757	.00106270
902	813604	30.0333	.00110865	942	887364	30.6920	.00106157
903	815409	30.0500	.00110742	943	889249	30.7083	.00106045
904	817216	30.0666	.00110619	944	891136	30.7246	.00105932
905	819025	30.0832	.00110497	945	893025	30.7409	.00105820
906	820836	30.0998	.00110375	946	894916	30.7571	.00105708
907	822649	30.1164	.00110254	947	896809	30.7734	.00105597
908	824464	30.1330	.00110132	948	898704	30.7896	.00105485
909	826281	30.1496	.00110011	949	900601	30.8058	.00105374
910	828100	30.1662	.00109890	950	902500	30.8221	.00105263
911	829921	30.1828	.00109769	951	904401	30.8383	.00105152
912	831744	30.1993	.00109649	952	906304	30.8545	.00105042
913	833569	30.2159	.00109529	953	908209	30.8707	.00104932
914	835396	30.2324	.00109409	954	910116	30.8869	.00104822
915	837225	30.2490	.00109290	955	912025	30.9031	.00104712
916	839056	30.2655	.00109170	956	913936	30.9192	.00104603
917	840889	30.2820	.00109051	957	915849	30.9354	.00104493
918	842724	30.2985	.00108932	958	917764	30.9516	.00104384
919	844561	30.3150	.00108814	959	919681	30.9677	.00104275
920	846400	30.3315	.00108696	960	921600	30.9839	.00104167

TABLE A
(Continued)

N	N²	√N	1/N	N	N²	√N	1/N
961	923521	31.0000	.00104058	981	962361	31.3209	.00101937
962	925444	31.0161	.00103950	982	964324	31.3369	.00101833
963	927369	31.0322	.00103842	983	966289	31.3528	.00101729
964	929296	31.0483	.00103734	984	968256	31.3688	.00101626
965	931225	31.0644	.00103627	985	970225	31.3847	.00101523
966	933156	31.0805	.00103520	986	972196	31.4006	.00101420
967	935089	31.0966	.00103413	987	974169	31.4166	.00101317
968	937024	31.1127	.00103306	988	976144	31.4325	.00101215
969	938961	31.1288	.00103199	989	978121	31.4484	.00101112
970	940900	31.1448	.00103093	990	980100	31.4643	.00101010
971	942841	31.1609	.00102987	991	982081	31.4802	.00100908
972	944784	31.1769	.00102881	992	984064	31.4960	.00100806
973	946729	31.1929	.00102775	993	986049	31.5119	.00100705
974	948676	31.2090	.00102669	994	988036	31.5278	.00100604
975	950625	31.2250	.00102564	995	990025	31.5436	.00100503
976	952576	31.2410	.00102549	996	992016	31.5595	.00100402
977	954529	31.2570	.00102354	997	994009	31.5753	.00100301
978	956484	31.2730	.00102249	998	996004	31.5911	.00100200
979	958441	31.2890	.00102145	999	998001	31.6070	.00100100
980	960400	31.3050	.00102041	1000	1000000	31.6228	.00100000

TABLE B

Percent area under the normal curve between the mean and z

z	.00	.01	.02	.03	.04	.05	.06	.07	.08	.09
0.0	00.00	00.40	00.80	01.20	01.60	01.99	02.39	02.79	03.19	03.59
0.1	03.98	04.38	04.78	05.17	05.57	05.96	06.36	06.75	07.14	07.53
0.2	07.93	08.32	08.71	09.10	09.48	09.87	10.26	10.64	11.03	11.41
0.3	11.79	12.17	12.55	12.93	13.31	13.68	14.06	14.43	14.80	15.17
0.4	15.54	15.91	16.28	16.64	17.00	17.36	17.72	18.08	18.44	18.79
0.5	19.15	19.50	19.85	20.19	20.54	20.88	21.23	21.57	21.90	22.24
0.6	22.57	22.91	23.24	23.57	23.89	24.22	24.54	24.86	25.17	25.49
0.7	25.80	26.11	26.42	26.73	27.04	27.34	27.64	27.94	28.23	28.52
0.8	28.81	29.10	29.39	29.67	29.95	30.23	30.51	30.78	31.06	31.33
0.9	31.59	31.86	32.12	32.38	32.64	32.89	33.15	33.40	33.65	33.89
1.0	34.13	34.38	34.61	34.85	35.08	35.31	35.54	35.77	35.99	36.21
1.1	36.43	36.65	36.86	37.08	37.29	37.49	37.70	37.90	38.10	38.30
1.2	38.49	38.69	38.88	39.07	39.25	39.44	39.62	39.80	39.97	40.15
1.3	40.32	40.49	40.66	40.82	40.99	41.15	41.31	41.47	41.62	41.77
1.4	41.92	42.07	42.22	42.36	42.51	42.65	42.79	42.92	43.06	43.19
1.5	43.32	43.45	43.57	43.70	43.82	43.94	44.06	44.18	44.29	44.41
1.6	44.52	44.63	44.74	44.84	44.95	45.05	45.15	45.25	45.35	45.45
1.7	45.54	45.64	45.73	45.82	45.91	45.99	46.08	46.16	46.25	46.33
1.8	46.41	46.49	46.56	46.64	46.71	46.78	46.86	46.93	46.99	47.06
1.9	47.13	47.19	47.26	47.32	47.38	47.44	47.50	47.56	47.61	47.67
2.0	47.72	47.78	47.83	47.88	47.93	47.98	48.03	48.08	48.12	48.17
2.1	48.21	48.26	48.30	48.34	48.38	48.42	48.46	48.50	48.54	48.57
2.2	48.61	48.64	48.68	48.71	48.75	48.78	48.81	48.84	48.87	48.90
2.3	48.93	48.96	48.98	49.01	49.04	49.06	49.09	49.11	49.13	49.16
2.4	49.18	49.20	49.22	49.25	49.27	49.29	49.31	49.32	49.34	49.36
2.5	49.38	49.40	49.41	49.43	49.45	49.46	49.48	49.49	49.51	49.52
2.6	49.53	49.55	49.56	49.57	49.59	49.60	49.61	49.62	49.63	49.64
2.7	49.65	49.66	49.67	49.68	49.69	49.70	49.71	49.72	49.73	49.74
2.8	49.74	49.75	49.76	49.77	49.77	49.78	49.79	49.79	49.80	49.81
2.9	49.81	49.82	49.82	49.83	49.84	49.84	49.85	49.85	49.86	49.86
3.0	49.87									
3.5	49.98									
4.0	49.997									
5.0	49.99997									

Source: B. F. Lindquist, *A first course in statistics*, 2nd ed. (New York: Houghton Mifflin, 1942). Reproduced by permission.

TABLE C

Critical values of t

	Level of significance for one-tailed test					
	.10	.05	.025	.01	.005	.0005
	Level of significance for two-tailed test					
df	.20	.10	.05	.02	.01	.001
1	3.078	6.314	12.706	31.821	63.657	636.619
2	1.886	2.920	4.303	6.965	9.925	31.598
3	1.638	2.353	3.182	4.541	5.841	12.941
4	1.533	2.132	2.776	3.747	4.604	8.610
5	1.476	2.015	2.571	3.365	4.032	6.859
6	1.440	1.943	2.447	3.143	3.707	5.959
7	1.415	1.895	2.365	2.998	3.449	5.405
8	1.397	1.860	2.306	2.896	3.355	5.041
9	1.383	1.833	2.262	2.821	3.250	4.781
10	1.372	1.812	2.228	2.764	3.169	4.587
11	1.363	1.796	2.201	2.718	3.106	4.437
12	1.356	1.782	2.179	2.681	3.055	4.318
13	1.350	1.771	2.160	2.650	3.012	4.221
14	1.345	1.761	2.145	2.624	2.977	4.140
15	1.341	1.753	2.131	2.602	2.947	4.073
16	1.337	1.746	2.120	2.583	2.921	4.015
17	1.333	1.740	2.110	2.567	2.898	3.965
18	1.330	1.734	2.101	2.552	2.878	3.922
19	1.328	1.729	2.093	2.539	2.861	3.883
20	1.325	1.725	2.086	2.528	2.845	3.850
21	1.323	1.721	2.080	2.518	2.831	3.819
22	1.321	1.717	2.074	2.508	2.819	3.792
23	1.319	1.714	2.069	2.500	2.807	3.767
24	1.318	1.711	2.064	2.492	2.797	3.745
25	1.316	1.708	2.060	2.485	2.787	3.725
26	1.315	1.706	2.056	2.479	2.779	3.707
27	1.314	1.703	2.052	2.473	2.771	3.690
28	1.313	1.701	2.048	2.467	2.763	3.674
29	1.311	1.699	2.045	2.462	2.756	3.659
30	1.310	1.697	2.042	2.457	2.750	3.646
40	1.303	1.684	2.021	2.423	2.704	3.551
60	1.296	1.671	2.000	2.390	2.660	3.460
120	1.289	1.658	1.980	2.358	2.617	3.373
∞	1.282	1.645	1.960	2.326	2.576	3.291

Source: Taken from Table III of R. A. Fisher and F. Yates, *Statistical tables for biological, agricultural and medical research*, 6th ed. (Edinburgh: Oliver and Boyd, 1963). Reproduced by permission of the authors and publishers.

TABLE D

Critical values of the Pearson r

df (=N − 2; N = number of pairs)	Level of significance for one-tailed test			
	.05	.025	.01	.005
	Level of significance for two-tailed test			
	.10	.05	.02	.01
1	.988	.997	.9995	.9999
2	.900	.950	.980	.990
3	.805	.878	.934	.959
4	.729	.811	.882	.917
5	.669	.754	.833	.874
6	.622	.707	.789	.834
7	.582	.666	.750	.798
8	.549	.632	.716	.765
9	.521	.602	.685	.735
10	.497	.576	.658	.708
11	.476	.553	.634	.684
12	.458	.532	.612	.661
13	.441	.514	.592	.641
14	.426	.497	.574	.623
15	.412	.482	.558	.606
16	.400	.468	.542	.590
17	.389	.456	.528	.575
18	.378	.444	.516	.561
19	.369	.433	.503	.549
20	.360	.423	.492	.537
21	.352	.413	.482	.526
22	.344	.404	.472	.515
23	.337	.396	.462	.505
24	.330	.388	.453	.496
25	.323	.381	.445	.487
26	.317	.374	.437	.479
27	.311	.367	.430	.471
28	.306	.361	.423	.463
29	.301	.355	.416	.456
30	.296	.349	.409	.449
35	.275	.325	.381	.418
40	.257	.304	.358	.393
45	.243	.288	.338	.372
50	.231	.273	.322	.354
60	.211	.250	.295	.325
70	.195	.232	.274	.302
80	.183	.217	.256	.283
90	.173	.205	.242	.267
100	.164	.195	.230	.254

Source: R. A. Fisher and F. Yates, Statistical tables for biological, agricultural and medical research, 6th ed. (Edinburgh: Oliver and Boyd, 1963). Reproduced by permission of the authors and publishers.

TABLE E

Critical values of r_s (Spearman rank-order correlation coefficient

No. of pairs (N)	Level of significance for one-tailed test			
	.05	.025	.01	.005
	Level of significance for two-tailed test			
	.10	.05	.02	.01
5	.900	1.000	1.000	—
6	.829	.886	.943	1.000
7	.714	.786	.893	.929
8	.643	.738	.833	.881
9	.600	.683	.783	.833
10	.564	.648	.746	.794
12	.506	.591	.712	.777
14	.456	.544	.645	.715
16	.425	.506	.601	.665
18	.399	.475	.564	.625
20	.377	.450	.534	.591
22	.359	.428	.508	.562
24	.343	.409	.485	.537
26	.329	.392	.465	.515
28	.317	.377	.448	.496
30	.306	.364	.432	.478

Source: G. E. Olds, Ann. Math. Statistics 9 (1938); 20 (1949). Reproduced by permission of the publisher.

TABLE F

Critical values of F ($\alpha = .05$ in lightface type, $\alpha = .01$ in boldface)

n_2	n_1 degrees of freedom (for numerator mean square)											
	1	2	3	4	5	6	7	8	9	10	11	12
1	161	200	216	225	230	234	237	239	241	242	243	244
	4,052	**4,999**	**5,403**	**5,625**	**5,764**	**5,859**	**5,928**	**5,981**	**6,022**	**6,056**	**6,082**	**6,106**
2	18.51	19.00	19.16	19.25	19.30	19.33	19.36	19.37	19.38	19.39	19.40	19.41
	98.49	**99.00**	**99.17**	**99.25**	**99.30**	**99.33**	**99.34**	**99.36**	**99.38**	**99.40**	**99.41**	**99.42**
3	10.13	9.55	9.28	9.12	9.01	8.94	8.88	8.84	8.81	8.78	8.76	8.74
	34.12	**30.82**	**29.46**	**28.71**	**28.24**	**27.91**	**27.67**	**27.49**	**27.34**	**27.23**	**27.13**	**27.05**
4	7.71	6.94	6.59	6.39	6.26	6.16	6.09	6.04	6.00	5.96	5.93	5.91
	21.20	**18.00**	**16.69**	**15.98**	**15.52**	**15.21**	**14.98**	**14.80**	**14.66**	**14.54**	**14.45**	**14.37**
5	6.61	5.79	5.41	5.19	5.05	4.95	4.88	4.82	4.78	4.74	4.70	4.68
	16.26	**13.27**	**12.06**	**11.39**	**10.97**	**10.67**	**10.45**	**10.27**	**10.15**	**10.05**	**9.96**	**9.89**
6	5.99	5.14	4.76	4.53	4.39	4.28	4.21	4.15	4.10	4.06	4.03	4.00
	13.74	**10.92**	**9.78**	**9.15**	**8.75**	**8.47**	**8.26**	**8.10**	**7.98**	**7.87**	**7.79**	**7.72**
7	5.59	4.74	4.35	4.12	3.97	3.87	3.79	3.73	3.68	3.63	3.60	3.57
	12.25	**9.55**	**8.45**	**7.85**	**7.46**	**7.19**	**7.00**	**6.84**	**6.71**	**6.62**	**6.54**	**6.47**
8	5.32	4.46	4.07	3.84	3.69	3.58	3.50	3.44	3.39	3.34	3.31	3.28
	11.26	**8.65**	**7.59**	**7.01**	**6.63**	**6.37**	**6.19**	**6.03**	**5.91**	**5.82**	**5.74**	**5.67**
9	5.12	4.26	3.86	3.63	3.48	3.37	3.29	3.23	3.18	3.13	3.10	3.07
	10.56	**8.02**	**6.99**	**6.42**	**6.06**	**5.80**	**5.62**	**5.47**	**5.35**	**5.26**	**5.18**	**5.11**
10	4.96	4.10	3.71	3.48	3.33	3.22	3.14	3.07	3.02	2.97	2.94	2.91
	10.04	**7.56**	**6.55**	**5.99**	**5.64**	**5.39**	**5.21**	**5.06**	**4.95**	**4.85**	**4.78**	**4.71**
11	4.84	3.98	3.59	3.36	3.20	3.09	3.01	2.95	2.90	2.86	2.82	2.79
	9.65	**7.20**	**6.22**	**5.67**	**5.32**	**5.07**	**4.88**	**4.74**	**4.63**	**4.54**	**4.46**	**4.40**
12	4.75	3.88	3.49	3.26	3.11	3.00	2.92	2.85	2.80	2.76	2.72	2.69
	9.33	**6.93**	**5.95**	**5.41**	**5.06**	**4.82**	**4.65**	**4.50**	**4.39**	**4.30**	**4.22**	**4.16**
13	4.67	3.80	3.41	3.18	3.02	2.92	2.84	2.77	2.72	2.67	2.63	2.60
	9.07	**6.70**	**5.74**	**5.20**	**4.86**	**4.62**	**4.44**	**4.30**	**4.19**	**4.10**	**4.02**	**3.96**

Source: Reprinted by permission from *Statistical methods*, 6th ed., by G. W. Snedecor and W. C. Cochran. © 1967 by the Iowa State University Press, Ames, Iowa.

TABLE F
Critical values of F ($\alpha = .05$ in lightface type, $\alpha = .01$ in boldface)

	n_1 degrees of freedom (for numerator mean square)											
14	16	20	24	30	40	50	75	100	200	500	∞
245	246	248	249	250	251	252	253	253	254	254	254
6,142	**6,169**	**6,208**	**6,234**	**6,258**	**6,286**	**6,302**	**6,323**	**6,334**	**6,352**	**6,361**	**6,366**
19.42	19.43	19.44	19.45	19.46	19.47	19.47	19.48	19.49	19.49	19.50	19.50
99.43	**99.44**	**99.45**	**99.46**	**99.47**	**99.48**	**99.48**	**99.49**	**99.49**	**99.49**	**99.50**	**99.50**
8.71	8.69	8.66	8.64	8.62	8.60	8.58	8.57	8.56	8.54	8.54	8.53
26.92	**26.83**	**26.69**	**26.60**	**26.50**	**26.41**	**26.35**	**26.27**	**26.23**	**26.18**	**26.14**	**26.12**
5.87	5.84	5.80	5.77	5.74	5.71	5.70	5.68	5.66	5.65	5.64	5.63
14.24	**14.15**	**14.02**	**13.93**	**13.83**	**13.74**	**13.69**	**13.61**	**13.57**	**13.52**	**13.48**	**13.46**
4.64	4.60	4.56	4.53	4.50	4.46	4.44	4.42	4.40	4.38	4.37	4.36
9.77	**9.68**	**9.55**	**9.47**	**9.38**	**9.29**	**9.24**	**9.17**	**9.13**	**9.07**	**9.04**	**9.02**
3.96	3.92	3.87	3.84	3.81	3.77	3.75	3.72	3.71	3.69	3.68	3.67
7.60	**7.52**	**7.39**	**7.31**	**7.23**	**7.14**	**7.09**	**7.02**	**6.99**	**6.94**	**6.90**	**6.88**
3.52	3.49	3.44	3.41	3.38	3.34	3.32	3.29	3.28	3.25	3.24	3.23
6.35	**6.27**	**6.15**	**6.07**	**5.98**	**5.90**	**5.85**	**5.78**	**5.75**	**5.70**	**5.67**	**5.65**
3.23	3.20	3.15	3.12	3.08	3.05	3.03	3.00	2.98	2.96	2.94	2.93
5.56	**5.48**	**5.36**	**5.28**	**5.20**	**5.11**	**5.06**	**5.00**	**4.96**	**4.91**	**4.88**	**4.86**
3.02	2.98	2.93	2.90	2.86	2.82	2.80	2.77	2.76	2.73	2.72	2.71
5.00	**4.92**	**4.80**	**4.73**	**4.64**	**4.56**	**4.51**	**4.45**	**4.41**	**4.36**	**4.33**	**4.31**
2.86	2.82	2.77	2.74	2.70	2.67	2.64	2.61	2.59	2.56	2.55	2.54
4.60	**4.52**	**4.41**	**4.33**	**4.25**	**4.17**	**4.12**	**4.05**	**4.01**	**3.96**	**3.93**	**3.91**
2.74	2.70	2.65	2.61	2.57	2.53	2.50	2.47	2.45	2.42	2.41	2.40
4.29	**4.21**	**4.10**	**4.02**	**3.94**	**3.86**	**3.80**	**3.74**	**3.70**	**3.66**	**3.62**	**3.60**
2.64	2.60	2.54	2.50	2.46	2.42	2.40	2.36	2.35	2.32	2.31	2.30
4.05	**3.98**	**3.86**	**3.78**	**3.70**	**3.61**	**3.56**	**3.49**	**3.46**	**3.41**	**3.38**	**3.36**
2.55	2.51	2.46	2.42	2.38	2.34	2.32	2.28	2.26	2.24	2.22	2.21
3.85	**3.78**	**3.67**	**3.59**	**3.51**	**3.42**	**3.37**	**3.30**	**3.27**	**3.21**	**3.18**	**3.16**

TABLE F
(Continued)

n_2	n_1 degrees of freedom (for numerator mean square)											
	1	2	3	4	5	6	7	8	9	10	11	12
14	4.60	3.74	3.34	3.11	2.96	2.85	2.77	2.70	2.65	2.60	2.56	2.53
	8.86	6.51	5.56	5.03	4.69	4.46	4.28	4.14	4.03	3.94	3.86	3.80
15	4.54	3.68	3.29	3.06	2.90	2.79	2.70	2.64	2.59	2.55	2.51	2.48
	8.68	6.36	5.42	4.89	4.56	4.32	4.14	4.00	3.89	3.80	3.73	3.67
16	4.49	3.63	3.24	3.01	2.85	2.74	2.66	2.59	2.54	2.49	2.45	2.42
	8.53	6.23	5.29	4.77	4.44	4.20	4.03	3.89	3.78	3.69	3.61	3.55
17	4.45	3.59	3.20	2.96	2.81	2.70	2.62	2.55	2.50	2.45	2.41	2.38
	8.40	6.11	5.18	4.67	4.34	4.10	3.93	3.79	3.68	3.59	3.52	3.45
18	4.41	3.55	3.16	2.93	2.77	2.66	2.58	2.51	2.46	2.41	2.37	2.34
	8.28	6.01	5.09	4.58	4.25	4.01	3.85	3.71	3.60	3.51	3.44	3.37
19	4.38	3.52	3.13	2.90	2.74	2.63	2.55	2.48	2.43	2.38	2.34	2.31
	8.18	5.93	5.01	4.50	4.17	3.94	3.77	3.63	3.52	3.43	3.36	3.30
20	4.35	3.49	3.10	2.87	2.71	2.60	2.52	2.45	2.40	2.35	2.31	2.28
	8.10	5.85	4.94	4.43	4.10	3.87	3.71	3.56	3.45	3.37	3.30	3.23
21	4.32	3.47	3.07	2.84	2.68	2.57	2.49	2.42	2.37	2.32	2.28	2.25
	8.02	5.78	4.87	4.37	4.04	3.81	3.65	3.51	3.40	3.31	3.24	3.17
22	4.30	3.44	3.05	2.82	2.66	2.55	2.47	2.40	2.35	2.30	2.26	2.23
	7.94	5.72	4.82	4.31	3.99	3.76	3.59	3.45	3.35	3.26	3.18	3.12
23	4.28	3.42	3.03	2.80	2.64	2.53	2.45	2.38	2.32	2.28	2.24	2.20
	7.88	5.66	4.76	4.26	3.94	3.71	3.54	3.41	3.30	3.21	3.14	3.07
24	4.26	3.40	3.01	2.78	2.62	2.51	2.43	2.36	2.30	2.26	2.22	2.18
	7.82	5.61	4.72	4.22	3.90	3.67	3.50	3.36	3.25	3.17	3.09	3.03
25	4.24	3.38	2.99	2.76	2.60	2.49	2.41	2.34	2.28	2.24	2.20	2.16
	7.77	5.57	4.68	4.18	3.86	3.63	3.46	3.32	3.21	3.13	3.05	2.99
26	4.22	3.37	2.98	2.74	2.59	2.47	2.39	2.32	2.27	2.22	2.18	2.15
	7.72	5.53	4.64	4.14	3.82	3.59	3.42	3.29	3.17	3.09	3.02	2.96

TABLE F

(Continued)

					n_1 degrees of freedom (for numerator mean square)							
14	16	20	24	30	40	50	75	100	200	500	∞
2.48	2.44	2.39	2.35	2.31	2.27	2.24	2.21	2.19	2.16	2.14	2.13
3.70	**3.62**	**3.51**	**3.43**	**3.34**	**3.26**	**3.21**	**3.14**	**3.11**	**3.06**	**3.02**	**3.00**
2.43	2.39	2.33	2.29	2.25	2.21	2.18	2.15	2.12	2.10	2.08	2.07
3.56	**3.48**	**3.36**	**3.29**	**3.20**	**3.12**	**3.07**	**3.00**	**2.97**	**2.92**	**2.89**	**2.87**
2.37	2.33	2.28	2.24	2.20	2.16	2.13	2.09	2.07	2.04	2.02	2.01
3.45	**3.37**	**3.25**	**3.18**	**3.10**	**3.01**	**2.96**	**2.89**	**2.86**	**2.80**	**2.77**	**2.75**
2.33	2.29	2.23	2.19	2.15	2.11	2.08	2.04	2.02	1.99	1.97	1.96
3.35	**3.27**	**3.16**	**3.08**	**3.00**	**2.92**	**2.86**	**2.79**	**2.76**	**2.70**	**2.67**	**2.65**
2.29	2.25	2.19	2.15	2.11	2.07	2.04	2.00	1.98	1.95	1.93	1.92
3.27	**3.19**	**3.07**	**3.00**	**2.91**	**2.83**	**2.78**	**2.71**	**2.68**	**2.62**	**2.59**	**2.57**
2.26	2.21	2.15	2.11	2.07	2.02	2.00	1.96	1.94	1.91	1.90	1.88
3.19	**3.12**	**3.00**	**2.92**	**2.84**	**2.76**	**2.70**	**2.63**	**2.60**	**2.54**	**2.51**	**2.49**
2.23	2.18	2.12	2.08	2.04	1.99	1.96	1.92	1.90	1.87	1.85	1.84
3.13	**3.05**	**2.94**	**2.86**	**2.77**	**2.69**	**2.63**	**2.56**	**2.53**	**2.47**	**2.44**	**2.42**
2.20	2.15	2.09	2.05	2.00	1.96	1.93	1.89	1.87	1.84	1.82	1.81
3.07	**2.99**	**2.88**	**2.80**	**2.72**	**2.63**	**2.58**	**2.51**	**2.47**	**2.42**	**2.38**	**2.36**
2.18	2.13	2.07	2.03	1.98	1.93	1.91	1.87	1.84	1.81	1.80	1.78
3.02	**2.94**	**2.83**	**2.75**	**2.67**	**2.58**	**2.53**	**2.46**	**2.42**	**2.37**	**2.33**	**2.31**
2.14	2.10	2.04	2.00	1.96	1.91	1.88	1.84	1.82	1.79	1.77	1.76
2.97	**2.89**	**2.78**	**2.70**	**2.62**	**2.53**	**2.48**	**2.41**	**2.37**	**2.32**	**2.28**	**2.26**
2.13	2.09	2.02	1.98	1.94	1.89	1.86	1.82	1.80	1.76	1.74	1.73
2.93	**2.85**	**2.74**	**2.66**	**2.58**	**2.49**	**2.44**	**2.36**	**2.33**	**2.27**	**2.23**	**2.21**
2.11	2.06	2.00	1.96	1.92	1.87	1.84	1.80	1.77	1.74	1.72	1.71
2.89	**2.81**	**2.70**	**2.62**	**2.54**	**2.45**	**2.40**	**2.32**	**2.29**	**2.23**	**2.19**	**2.17**
2.10	2.05	1.99	1.95	1.90	1.85	1.82	1.78	1.76	1.72	1.70	1.69
2.86	**2.77**	**2.66**	**2.58**	**2.50**	**2.41**	**2.36**	**2.28**	**2.25**	**2.19**	**2.15**	**2.13**

TABLE F
(Continued)

n_2	n_1 degrees of freedom (for numerator mean square)											
	1	2	3	4	5	6	7	8	9	10	11	12
27	4.21	3.35	2.96	2.73	2.57	2.46	2.37	2.30	2.25	2.20	2.16	2.13
	7.68	**5.49**	**4.60**	**4.11**	**3.79**	**3.56**	**3.39**	**3.26**	**3.14**	**3.06**	**2.98**	**2.93**
28	4.20	3.34	2.95	2.71	2.56	2.44	2.36	2.29	2.24	2.19	2.15	2.12
	7.64	**5.45**	**4.57**	**4.07**	**3.76**	**3.53**	**3.36**	**3.23**	**3.11**	**3.03**	**2.95**	**2.90**
29	4.18	3.33	2.93	2.70	2.54	2.43	2.35	2.28	2.22	2.18	2.14	2.10
	7.60	**5.42**	**4.54**	**4.04**	**3.73**	**3.50**	**3.33**	**3.20**	**3.08**	**3.00**	**2.92**	**2.87**
30	4.17	3.32	2.92	2.69	2.53	2.42	2.34	2.27	2.21	2.16	2.12	2.09
	7.56	**5.39**	**4.51**	**4.02**	**3.70**	**3.47**	**3.30**	**3.17**	**3.06**	**2.98**	**2.90**	**2.84**
32	4.15	3.30	2.90	2.67	2.51	2.40	2.32	2.25	2.19	2.14	2.10	2.07
	7.50	**5.34**	**4.46**	**3.97**	**3.66**	**3.42**	**3.25**	**3.12**	**3.01**	**2.94**	**2.86**	**2.80**
34	4.13	3.28	2.88	2.65	2.49	2.38	2.30	2.23	2.17	2.12	2.08	2.05
	7.44	**5.29**	**4.42**	**3.93**	**3.61**	**3.38**	**3.21**	**3.08**	**2.97**	**2.89**	**2.82**	**2.76**
36	4.11	3.26	2.86	2.63	2.48	2.36	2.28	2.21	2.15	2.10	2.06	2.03
	7.39	**5.25**	**4.38**	**3.89**	**3.58**	**3.35**	**3.18**	**3.04**	**2.94**	**2.86**	**2.78**	**2.72**
38	4.10	3.25	2.85	2.62	2.46	2.35	2.26	2.19	2.14	2.09	2.05	2.02
	7.35	**5.21**	**4.34**	**3.86**	**3.54**	**3.32**	**3.15**	**3.02**	**2.91**	**2.82**	**2.75**	**2.69**
40	4.08	3.23	2.84	2.61	2.45	2.34	2.25	2.18	2.12	2.07	2.04	2.00
	7.31	**5.18**	**4.31**	**3.83**	**3.51**	**3.29**	**3.12**	**2.99**	**2.88**	**2.80**	**2.73**	**2.66**
42	4.07	3.22	2.83	2.59	2.44	2.32	2.24	2.17	2.11	2.06	2.02	1.99
	7.27	**5.15**	**4.29**	**3.80**	**3.49**	**3.26**	**3.10**	**2.96**	**2.86**	**2.77**	**2.70**	**2.64**
44	4.06	3.21	2.82	2.58	2.43	2.31	2.23	2.16	2.10	2.05	2.01	1.98
	7.24	**5.12**	**4.26**	**3.78**	**3.46**	**3.24**	**3.07**	**2.94**	**2.84**	**2.75**	**2.68**	**2.62**
46	4.05	3.20	2.81	2.57	2.42	2.30	2.22	2.14	2.09	2.04	2.00	1.97
	7.21	**5.10**	**4.24**	**3.76**	**3.44**	**3.22**	**3.05**	**2.92**	**2.82**	**2.73**	**2.66**	**2.60**
48	4.04	3.19	2.80	2.56	2.41	2.30	2.21	2.14	2.08	2.03	1.99	1.96
	7.19	**5.08**	**4.22**	**3.74**	**3.42**	**3.20**	**3.04**	**2.90**	**2.80**	**2.71**	**2.64**	**2.58**

TABLE F
(Continued)

	n_1 degrees of freedom (for numerator mean square)											
14	16	20	24	30	40	50	75	100	200	500	∞
2.08	2.03	1.97	1.93	1.88	1.84	1.80	1.76	1.74	1.71	1.68	1.67
2.83	**2.74**	**2.63**	**2.55**	**2.47**	**2.38**	**2.33**	**2.25**	**2.21**	**2.16**	**2.12**	**2.10**
2.06	2.02	1.96	1.91	1.87	1.81	1.78	1.75	1.72	1.69	1.67	1.65
2.80	**2.71**	**2.60**	**2.52**	**2.44**	**2.35**	**2.30**	**2.22**	**2.18**	**2.13**	**2.09**	**2.06**
2.05	2.00	1.94	1.90	1.85	1.80	1.77	1.73	1.71	1.68	1.65	1.64
2.77	**2.68**	**2.57**	**2.49**	**2.41**	**2.32**	**2.27**	**2.19**	**2.15**	**2.10**	**2.06**	**2.03**
2.04	1.99	1.93	1.89	1.84	1.79	1.76	1.72	1.69	1.66	1.64	1.62
2.74	**2.66**	**2.55**	**2.47**	**2.38**	**2.29**	**2.24**	**2.16**	**2.13**	**2.07**	**2.03**	**2.01**
2.02	1.97	1.91	1.86	1.82	1.76	1.74	1.69	1.67	1.64	1.61	1.59
2.70	**2.62**	**2.51**	**2.42**	**2.34**	**2.25**	**2.20**	**2.12**	**2.08**	**2.02**	**1.98**	**1.96**
2.00	1.95	1.89	1.84	1.80	1.74	1.71	1.67	1.64	1.61	1.59	1.57
2.66	**2.58**	**2.47**	**2.38**	**2.30**	**2.21**	**2.15**	**2.08**	**2.04**	**1.98**	**1.94**	**1.91**
1.98	1.93	1.87	1.82	1.78	1.72	1.69	1.65	1.62	1.59	1.56	1.55
2.62	**2.54**	**2.43**	**2.35**	**2.26**	**2.17**	**2.12**	**2.04**	**2.00**	**1.94**	**1.90**	**1.87**
1.96	1.92	1.85	1.80	1.76	1.71	1.67	1.63	1.60	1.57	1.54	1.53
2.59	**2.51**	**2.40**	**2.32**	**2.22**	**2.14**	**2.08**	**2.00**	**1.97**	**1.90**	**1.86**	**1.84**
1.95	1.90	1.84	1.79	1.74	1.69	1.66	1.61	1.59	1.55	1.53	1.51
2.56	**2.49**	**2.37**	**2.29**	**2.20**	**2.11**	**2.05**	**1.97**	**1.94**	**1.88**	**1.84**	**1.81**
1.94	1.89	1.82	1.78	1.73	1.68	1.64	1.60	1.57	1.54	1.51	1.49
2.54	**2.46**	**2.35**	**2.26**	**2.17**	**2.08**	**2.02**	**1.94**	**1.91**	**1.85**	**1.80**	**1.78**
1.92	1.88	1.81	1.76	1.72	1.66	1.63	1.58	1.56	1.52	1.50	1.48
2.52	**2.44**	**2.32**	**2.24**	**2.15**	**2.06**	**2.00**	**1.92**	**1.88**	**1.82**	**1.78**	**1.75**
1.91	1.87	1.80	1.75	1.71	1.65	1.62	1.57	1.54	1.51	1.48	1.46
2.50	**2.42**	**2.30**	**2.22**	**2.13**	**2.04**	**1.98**	**1.90**	**1.86**	**1.80**	**1.76**	**1.72**
1.90	1.86	1.79	1.74	1.70	1.64	1.61	1.56	1.53	1.50	1.47	1.45
2.48	**2.40**	**2.28**	**2.20**	**2.11**	**2.02**	**1.96**	**1.88**	**1.84**	**1.78**	**1.73**	**1.70**

TABLE F
(Continued)

n_2	n_1 degrees of freedom (for numerator mean square)											
	1	2	3	4	5	6	7	8	9	10	11	12
50	4.03	3.18	2.79	2.56	2.40	2.29	2.20	2.13	2.07	2.02	1.98	1.95
	7.17	5.06	4.20	3.72	3.41	3.18	3.02	2.88	2.78	2.70	2.62	2.56
55	4.02	3.17	2.78	2.54	2.38	2.27	2.18	2.11	2.05	2.00	1.97	1.93
	7.12	5.01	4.16	3.68	3.37	3.15	2.98	2.85	2.75	2.66	2.59	2.53
60	4.00	3.15	2.76	2.52	2.37	2.25	2.17	2.10	2.04	1.99	1.95	1.92
	7.08	4.98	4.13	3.65	3.34	3.12	2.95	2.82	2.72	2.63	2.56	2.50
65	3.99	3.14	2.75	2.51	2.36	2.24	2.15	2.08	2.02	1.98	1.94	1.90
	7.04	4.95	4.10	3.62	3.31	3.09	2.93	2.79	2.70	2.61	2.54	2.47
70	3.98	3.13	2.74	2.50	2.35	2.23	2.14	2.07	2.01	1.97	1.93	1.89
	7.01	4.92	4.08	3.60	3.29	3.07	2.91	2.77	2.67	2.59	2.51	2.45
80	3.96	3.11	2.72	2.48	2.33	2.21	2.12	2.05	1.99	1.95	1.91	1.88
	6.96	4.88	4.04	3.56	3.25	3.04	2.87	2.74	2.64	2.55	2.48	2.41
100	3.94	3.09	2.70	2.46	2.30	2.19	2.10	2.03	1.97	1.92	1.88	1.85
	6.90	4.82	3.98	3.51	3.20	2.99	2.82	2.69	2.59	2.51	2.43	2.36
125	3.92	3.07	2.68	2.44	2.29	2.17	2.08	2.01	1.95	1.90	1.86	1.83
	6.84	4.78	3.94	3.47	3.17	2.95	2.79	2.65	2.56	2.47	2.40	2.33
150	3.91	3.06	2.67	2.43	2.27	2.16	2.07	2.00	1.94	1.89	1.85	1.82
	6.81	4.75	3.91	3.44	3.14	2.92	2.76	2.62	2.53	2.44	2.37	2.30
200	3.89	3.04	2.65	2.41	2.26	2.14	2.05	1.98	1.92	1.87	1.83	1.80
	6.76	4.71	3.88	3.41	3.11	2.90	2.73	2.60	2.50	2.41	2.34	2.28
400	3.86	3.02	2.62	2.39	2.23	2.12	2.03	1.96	1.90	1.85	1.81	1.78
	6.70	4.66	3.83	3.36	3.06	2.85	2.69	2.55	2.46	2.37	2.29	2.23
1000	3.85	3.00	2.61	2.38	2.22	2.10	2.02	1.95	1.89	1.84	1.80	1.76
	6.66	4.62	3.80	3.34	3.04	2.82	2.66	2.53	2.43	2.34	2.26	2.20
∞	3.84	2.99	2.60	2.37	2.21	2.09	2.01	1.94	1.88	1.83	1.79	1.75
	6.64	4.60	3.78	3.32	3.02	2.80	2.64	2.51	2.41	2.32	2.24	2.18

TABLE F

(Continued)

			n_1 degrees of freedom (for numerator mean square)								
14	16	20	24	30	40	50	75	100	200	500	∞
1.90	1.85	1.78	1.74	1.69	1.63	1.60	1.55	1.52	1.48	1.46	1.44
2.46	**2.39**	**2.26**	**2.18**	**2.10**	**2.00**	**1.94**	**1.86**	**1.82**	**1.76**	**1.71**	**1.68**
1.88	1.83	1.76	1.72	1.67	1.61	1.58	1.52	1.50	1.46	1.43	1.41
2.43	**2.35**	**2.23**	**2.15**	**2.06**	**1.96**	**1.90**	**1.82**	**1.78**	**1.71**	**1.66**	**1.64**
1.86	1.81	1.75	1.70	1.65	1.59	1.56	1.50	1.48	1.44	1.41	1.39
2.40	**2.32**	**2.20**	**2.12**	**2.03**	**1.93**	**1.87**	**1.79**	**1.74**	**1.68**	**1.63**	**1.60**
1.84	1.80	1.73	1.68	1.63	1.57	1.54	1.49	1.46	1.42	1.39	1.37
2.37	**2.30**	**2.18**	**2.09**	**2.00**	**1.90**	**1.84**	**1.76**	**1.71**	**1.64**	**1.60**	**1.56**
1.84	1.79	1.72	1.67	1.62	1.56	1.53	1.47	1.45	1.40	1.37	1.35
2.35	**2.28**	**2.15**	**2.07**	**1.98**	**1.88**	**1.82**	**1.74**	**1.69**	**1.62**	**1.56**	**1.53**
1.82	1.77	1.70	1.65	1.60	1.54	1.51	1.45	1.42	1.38	1.35	1.32
2.32	**2.24**	**2.11**	**2.03**	**1.94**	**1.84**	**1.78**	**1.70**	**1.65**	**1.57**	**1.52**	**1.49**
1.79	1.75	1.68	1.63	1.57	1.51	1.48	1.42	1.39	1.34	1.30	1.28
2.26	**2.19**	**2.06**	**1.98**	**1.89**	**1.79**	**1.73**	**1.64**	**1.59**	**1.51**	**1.46**	**1.43**
1.77	1.72	1.65	1.60	1.55	1.49	1.45	1.39	1.36	1.31	1.27	1.25
2.23	**2.15**	**2.03**	**1.94**	**1.85**	**1.75**	**1.68**	**1.59**	**1.54**	**1.46**	**1.40**	**1.37**
1.76	1.71	1.64	1.59	1.54	1.47	1.44	1.37	1.34	1.29	1.25	1.22
2.20	**2.12**	**2.00**	**1.91**	**1.83**	**1.72**	**1.66**	**1.56**	**1.51**	**1.43**	**1.37**	**1.33**
1.74	1.69	1.62	1.57	1.52	1.45	1.42	1.35	1.32	1.26	1.22	1.19
2.17	**2.09**	**1.97**	**1.88**	**1.79**	**1.69**	**1.62**	**1.53**	**1.48**	**1.39**	**1.33**	**1.28**
1.72	1.67	1.60	1.54	1.49	1.42	1.38	1.32	1.28	1.22	1.16	1.13
2.12	**2.04**	**1.92**	**1.84**	**1.74**	**1.65**	**1.57**	**1.47**	**1.42**	**1.32**	**1.24**	**1.19**
1.70	1.65	1.58	1.53	1.47	1.41	1.36	1.30	1.26	1.19	1.13	1.08
2.09	**2.01**	**1.89**	**1.81**	**1.71**	**1.61**	**1.54**	**1.44**	**1.38**	**1.28**	**1.19**	**1.11**
1.69	1.64	1.57	1.52	1.46	1.40	1.35	1.28	1.24	1.17	1.11	1.00
2.07	**1.99**	**1.87**	**1.79**	**1.69**	**1.59**	**1.52**	**1.41**	**1.36**	**1.25**	**1.15**	**1.00**

TABLE G

Critical values of chi-square

df*	.20	.10	.05	.02	.01	.001
			Level of significance			
1	1.64	2.71	3.84	5.41	6.63	10.83
2	3.22	4.61	5.99*	7.82	9.21	13.82
3	4.64	6.25	7.82	9.84	11.34	16.27
4	5.99	7.78	9.49	11.67	13.28	18.46
5	7.29	9.24	11.07	13.39	15.09	20.52
6	8.56	10.64	12.59	15.03	16.81	22.46
7	9.80	12.02	14.07	16.62	18.48	24.32
8	11.03	13.36	15.51	18.17	20.09	26.12
9	12.24	14.68	16.92	19.68	21.67	27.88
10	13.44	15.99	18.31	21.16	23.21	29.59
11	14.63	17.28	19.68	22.62	24.72	31.26
12	15.81	18.55	21.03	24.05	26.22	32.91
13	16.98	19.81	22.36	25.47	27.69	34.53
14	18.15	21.06	23.68	26.87	29.14	36.12
15	19.31	22.31	25.00	28.26	30.58	37.70
16	20.46	23.54	26.30	29.63	32.00	39.25
17	21.62	24.77	27.59	31.00	33.41	40.79
18	22.76	25.99	28.87	32.35	34.81	42.31
19	23.90	27.20	30.14	33.69	36.19	43.82
20	25.04	28.41	31.41	35.02	37.57	45.32
21	26.17	29.62	32.67	36.34	38.93	46.80
22	27.30	30.81	33.92	37.66	40.29	48.27
23	28.43	32.01	35.17	38.97	41.64	49.73
24	29.55	33.20	36.42	40.27	42.98	51.18
25	30.68	34.38	37.65	41.57	44.31	52.62
26	31.80	35.56	38.89	42.86	45.64	54.05
27	32.91	36.74	40.11	44.14	46.96	55.48
28	34.03	37.92	41.34	45.42	48.28	56.89
29	35.14	39.09	42.56	46.69	49.59	58.30
30	36.25	40.26	43.77	47.96	50.89	59.70

Source: R. A. Fisher and F. Yates, *Statistical tables for biological, agricultural and medical research*, 6th ed. (Edinburgh: Oliver and Boyd, 1963). Reproduced by permission of the authors and publishers.

* For *df* greater than 30, the value obtained from the expression $\sqrt{2\chi^2} - \sqrt{2df - 1}$ may be used as a *t* ratio.

TABLE H

Power as a function of δ and significance criterion (α)

	One-tailed test (α)					One-tailed test (α)			
	.05	.025	.01	.005		.05	.025	.01	.005
	Two-tailed test (α)					Two-tailed test (α)			
δ	.10	.05	.02	.01	δ	.10	.05	.02	.01
0.0	.10*	.05*	.02	.01	2.5	.80	.71	.57	.47
0.1	.10*	.05*	.02	.01	2.6	.83	.74	.61	.51
0.2	.11*	.05	.02	.01	2.7	.85	.77	.65	.55
0.3	.12*	.06	.03	.01	2.8	.88	.80	.68	.59
0.4	.13*	.07	.03	.01	2.9	.90	.83	.72	.63
0.5	.14	.08	.03	.02	3.0	.91	.85	.75	.66
0.6	.16	.09	.04	.02	3.1	.93	.87	.78	.70
0.7	.18	.11	.05	.03	3.2	.94	.89	.81	.73
0.8	.21	.13	.06	.04	3.3	.96	.91	.83	.77
0.9	.23	.15	.08	.05	3.4	.96	.93	.86	.80
1.0	.26	.17	.09	.06	3.5	.97	.94	.88	.82
1.1	.30	.20	.11	.07	3.6	.97	.95	.90	.85
1.2	.33	.22	.13	.08	3.7	.98	.96	.92	.87
1.3	.37	.26	.15	.10	3.8	.98	.97	.93	.89
1.4	.40	.29	.18	.12	3.9	.99	.97	.94	.91
1.5	.44	.32	.20	.14	4.0	.99	.98	.95	.92
1.6	.48	.36	.23	.16	4.1	.99	.98	.96	.94
1.7	.52	.40	.27	.19	4.2	.99	.99	.97	.95
1.8	.56	.44	.30	.22	4.3	†	.99	.98	.96
1.9	.60	.48	.33	.25	4.4		.99	.98	.97
2.0	.64	.52	.37	.28	4.5		.99	.99	.97
2.1	.68	.56	.41	.32	4.6		†	.99	.98
2.2	.71	.59	.45	.35	4.7			.99	.98
2.3	.74	.63	.49	.39	4.8			.99	.99
2.4	.77	.67	.53	.43	4.9			.99	.99
					5.0			†	.99
					5.1				.99
					5.2				†

* Values inaccurate for *one-tailed* test by more than .01.

† The power at and below this point is greater than .995.

TABLE I

δ as a function of significance criterion (α) and power

δ	One-tailed test (α)			
	.05	.025	.01	.005
	Two-tailed test (α)			
	.10	.05	.02	.01
.25	0.97	1.29	1.65	1.90
.50	1.64	1.96	2.33	2.58
.60	1.90	2.21	2.58	2.83
.67	2.08	2.39	2.76	3.01
.70	2.17	2.48	2.85	3.10
.75	2.32	2.63	3.00	3.25
.80	2.49	2.80	3.17	3.42
.85	2.68	3.00	3.36	3.61
.90	2.93	3.24	3.61	3.86
.95	3.29	3.60	3.97	4.22
.99	3.97	4.29	4.65	4.90
.999	4.37	5.05	5.42	5.67

Glossary of Terms

A priori **comparison**   In analysis of variance, a significance test between two group means that was planned before the results were in.

Absolute values   Numerical values with the signs ignored; thus all absolute values are treated as positive.

Alpha   The probability of committing a Type I error.

Alternative hypothesis   The hypothesis which states that the null hypothesis is *not true* and specifies some other value or set of values for the population parameter(s) in question.

Analysis of variance   Procedures for testing hypotheses about the equality of population means.

Area of rejection   All numerical results of a statistical test that will cause the null hypothesis to be rejected.

Attenuation   A synonym for restriction of range.

Beta   The probability of committing a Type II error.

Between-group variance   In analysis of variance, differences among the means of the various groups.

Bimodal distribution   A distribution that has two pronounced peaks when graphed as a frequency polygon.

Biserial correlation coefficient   A measure of the strength of the relationship between one dichotomous and one continuous variable, used when the distribution underlying the dichotomous variable is assumed to be normal.

Bivariate normal distribution   A joint distribution of two variables, wherein scores on one variable are normally distributed for each score value of the other.

Box-and-whisker plot   A pictorial description of the median, first and third quartiles, and lowest and highest scores of one or more groups.

Central tendency   The general location of a set of scores.

Chi square   The statistical model used to test hypotheses when data are in the form of frequencies.

Class interval   One set of score values in a grouped frequency distribution.

Confidence interval   A range of values within which a specified population parameter has a given probability of falling.

Confidence limits   The upper and lower end points of a confidence interval.

Constant   A numerical value that is exactly the same for all cases or subjects; the converse of *variable.*

Contingency coefficient   A measure of the strength of the relationship between two categorical variables.

Continuous variable   A variable for which it is theoretically possible for any value to occur between any specified pair of score values.

Control group   A group that does *not* receive the treatment whose effects are being investigated by the researcher. Used as a baseline against which to evaluate the performance of the experimental group.

Correlation coefficient   A measure of the extent to which scores on one variable are related to scores on a second variable.

Cramér's phi   A measure of the strength of the relationship between two variables, both of which are categorical.

Criterion   The variable being predicted in a linear regression analysis; a dependent variable.

Criterion of significance   A numerical value or decision rule that specifies when the null hypothesis is to be rejected.

Critical interval   When computing percentiles, the class interval in which the specified raw score or critical case number falls.

Cumulative frequency distribution   A listing that shows how many times a given score or less was obtained.

Decile   A transformed score that divides the number of cases into ten equal parts.

Degrees of freedom   The number of quantities that are free to vary when we estimate the value of a parameter from a statistic.

Delta   In power analysis, an index that combines the population effect size and the sample size.

Dependent variable   A variable that changes with changes in one or more independent variables.

Descriptive statistics   Mathematical procedures for summarizing and describing the characteristics of a set of data.

Deviation score   The difference between a score and the mean of the set of scores of which it is a member.

Dichotomous variable   A variable with only two categories.

Discrete variable   A variable for which it is *not* theoretically possible for any value to occur between any specified pair of score values.

Distribution-free statistical test   A statistical test that does *not* require any assumptions about the shape of the distribution in the population.

Effect size   How large the phenomenon we wish to investigate is in the population; the extent to which the null hypothesis is false.

Empirical sampling distribution   A synonym for experimental sampling distribution.

Epsilon   In one-way analysis of variance, a measure that estimates the strength of the relationship between the independent and dependent variables in the population.

Error variance   In analysis of variance, differences among scores or group means that *cannot* be explained by the experimental treatment(s).

Experimental group   A group that receives the treatment whose effects are being investigated by the researcher.

Experimental sampling distribution   A sample whose elements are statistics (for example, sample means) obtained by drawing repeated samples from the population and computing that statistic for each sample.

Experimentwise error rate   The rate of occurrence of *any* (one or more) Type I errors when a *series* of individual statistical tests is conducted.

***F* distributions**   The statistical model used to test hypotheses when the analysis involves the comparison of variance estimates.

Factorial design   A procedure used to study the relationship of two or more independent variables (factors) to a dependent variable.

Five-number summary   A procedure for summarizing important numerical characteristics of a set of data: the median, first and third quartile, and lowest and highest score.

Frequency   The number of times a specified score or range of scores was obtained.

Frequency polygon   A graph showing how often each score or range of scores was obtained.

Gamma   A synonym for effect size.

Generalize   Apply results obtained from a sample to a specific population.

Glass rank biserial correlation   A measure of the strength of the relationship between membership in one of two independent samples and a set of ranked data.

Grand median   In the median test, the median of all of the cases (all samples combined).

Grouped frequency distribution   A listing of sets of two or more score values (class intervals) together with the number of times that scores in each class interval were obtained.

Histogram   A bar graph showing how often each score or range of scores was obtained.

Homogeneity of variance   In analysis of variance, equality of the variances for all treatment populations.

Hypothesis testing   Procedures for deciding whether to retain or reject the null hypothesis about one or more population parameters.

Independent events   Events whose occurrence or nonoccurrence is unrelated to the occurrence (or probability of occurrence) of specified other events.

Independent samples   Samples such that any element in one sample has no connection of any kind with any element in another sample.

Independent variable   A variable whose variation is studied with regard to its effect on another (dependent) variable(s).

Inferential statistics   Mathematical procedures for drawing inferences about characteristics of a population, based on what is observed in a sample from that population.

Interaction   The joint effect of two or more independent variables on a dependent variable, over and above their separate (main) effects.

Interval estimate   A synonym for confidence interval.

Interval size   The number of scores in each class interval in a grouped frequency distribution.

J-curve   A distribution that looks like the letter J or its mirror image when graphed as a frequency polygon.

Kruskal-Wallis *H* test   A nonparametric procedure for testing hypotheses about differences among the locations of two or more independent populations, used when data are in the form of ranks.

Least squares regression line   The regression line that minimizes the sum of squared errors in prediction (the sum of squared deviations between the predicted scores and actual scores).

Linear regression   Procedures for determining the straight line that will enable us to predict scores on one variable (the criterion) from scores on another variable (the predictor), while minimizing the amount of (squared) error.

Lower real limit   The lower end point of a class interval in a grouped frequency distribution.

Marginal frequency   A column or row total in a two-way table of joint frequencies.

Matched-pairs rank biserial correlation   A measure of the strength of the relationship between membership in one of two matched samples and a set of ranked data.

Matched samples   Samples such that each element in one sample is paired with one element in the other sample.

Matched t test   Procedures for testing hypotheses about the difference between two population means, using matched samples.

Mean   A measure of the central tendency of a set of scores, obtained by summing all scores and dividing by the number of scores.

Mean-on-spoke representation   A graphic method that depicts the mean and standard deviation of one or more groups.

Mean square   In analysis of variance, an estimate of the population variance that is obtained by dividing a sum of squares by its associated degrees of freedom.

Median   The score such that exactly half the scores in the group are higher and exactly half the scores are equal or lower; the score corresponding to the 50th percentile.

Median test   A method for testing statistical significance when testing hypotheses about differences in location among two or more independent populations.

Midpoint   The value halfway between the lower real limit and the upper real limit of a class interval.

Mode   The score that occurs most often in a frequency distribution.

Multiple comparisons   In analysis of variance, procedures for testing hypotheses about differences between specified pairs of means.

Mutually exclusive events   Events that cannot happen simultaneously.

Negatively skewed distribution   A distribution with some extremely low scores, resulting in a frequency polygon with a pronounced tail at the left.

Nonparametric statistical test   A statistical test that does not involve the estimation of any population parameters.

Normal distribution   A particular bell-shaped, symmetric, and unimodal distribution defined by a specific mathematical equation.

Null hypothesis   The hypothesis that specifies the value of a population parameter or a difference between two or more population parameters (usually zero), and that is assumed to be true at the outset of the statistical analysis.

Odds against an event   The ratio of the number of unfavorable outcomes to the number of favorable outcomes.

One-tailed test of significance   A statistical test wherein the null hypothesis can only be rejected if results are in the direction predicted by the experimenter.

Parameter   A numerical quantity that summarizes some characteristic of a population.

Pearson r   A measure of the strength of the linear relationship between two variables, both of which are continuous.

Percentile   A score at or below which a given percent of the cases fall.

Percentile rank   The percent of cases in a specific reference group that fall at or below a given score.

Phi   A measure of the strength of the relationship between two variables, both of which are dichotomous (that is, have only two categories).

Point biserial correlation coefficient   A measure of the strength of the relationship between one dichotomous and one continuous variable.

Point estimate   A single statistic used to estimate the value of the corresponding parameter.

Pooled variance   In the significance test of the difference between two means, an estimate of the population variance (assumed to be the same for both populations) obtained from a weighted average of the sample variances.

Population   *All* of the cases in which a researcher is interested; a (usually very large) group of people, animals, objects, or responses that are alike in at least one respect.

Population variance estimate   An estimate of the variability in a population, obtained by dividing the sum of squared deviations from the sample mean by $N - 1$.

Positively skewed distribution   A distribution with some extremely high scores, resulting in a frequency polygon with a pronounced tail at the right.

Post hoc comparison   In analysis of variance, a significance test between two group means that was planned after the results were obtained.

Power efficiency of a nonparametric test   How many fewer cases, expressed as a percent, that are required by a parametric test to have the same power as the nonparametric test when the assumptions of the parametric test are met.

Power of a statistical test   The probability of obtaining a statistically significant result if the null hypothesis is actually false; the probability of rejecting a false null hypothesis.

Predicted score   In linear regression, the score on the criterion that is estimated for subjects with a specified score on the predictor.

Predictor   In linear regression, the variable from which criterion scores are estimated; an independent variable.

Probability of an event   The number of ways the specified event can occur, divided by the total number of possible events.

Protected t test   In analysis of variance, procedures for conducting multiple comparisons between pairs of group means while keeping the experimentwise error rate at a reasonable level.

Quartile   A transformed score that divides the number of cases into four equal parts.

Random sample   A sample drawn in such a way that each element in the population has an equal chance of being included in the sample.

Range   The largest score minus the smallest score.

Rank   A measure that shows where a given case falls with respect to others in the group, but gives no indication of the distance between the cases. Thus, ranks provide less information than do scores.

Rank-order correlation coefficient   A measure of the strength of the relationship between two variables, both of which are in the form of ranks.

Rank-sum test   A nonparametric procedure for testing hypotheses about the difference in location between two populations; used when data are in the form of ranks.

Raw score   A score not subjected to any statistical transformations, such as the number correct on a test.

Rectangular distribution   A distribution wherein each score occurs with the same frequency.

Regular frequency distribution   A listing of every score value, from the lowest to the highest, together with the number of times that each score was obtained (its frequency).

Restriction of range   Low variability, caused by the way in which the samples were defined, that will underestimate the correlation between that variable and other variables.

Robust statistical test   A statistical test that gives fairly accurate results even if the underlying assumptions are not met.

Sample   Any subgroup of cases drawn from a clearly specified population.

Sample size   The number of cases in a sample.

Sampling error   Differences between the value of a statistic observed in a sample and the corresponding population parameter (thus, error), caused by the accident of which cases happened to be included in the sample.

Scatter plot A graph showing the relationship between two variables, where each score is expressed as a point.

Sign test A method for testing statistical significance when testing hypotheses about differences in location between two matched populations.

Significance test A procedure for deciding whether to retain or reject a null hypothesis.

Skewed distribution A distribution with some extremely low scores (negatively skewed, skewed to the left) or some extremely high scores (positively skewed, skewed to the right), resulting in a frequency polygon with a pronounced tail in one direction.

Slope In linear regression, the average rate of change in the criterion per unit increase in the predictor.

Standard deviation A measure of the variability of the scores in a specified group, or how much the scores differ from one another; the positive square root of the variance.

Standard error The standard deviation of some statistic in a sampling distribution (rather than scores), which tells us how trustworthy the single statistic we have on hand is as an estimate of the corresponding parameter. Examples include a population mean *(standard error of the mean)*, a population proportion *(standard error of a proportion)*, and the difference between two population means *(standard error of the difference)*.

Standard error of estimate In linear regression, a measure of the variability (average error) in prediction.

Standard scores A synonym for *Z* scores.

Statistic A numerical quantity that summarizes some characteristic of a sample.

Statistical model A summary of the probability of all possible events of a specified type (for example, each possible value of a sample mean drawn at random from one population) under specified conditions (for example, that the null hypothesis is true). Examples include the normal distribution, the *t* distributions, the *F* distributions, and chi square.

Statistical significance Occurs when the results of a statistical analysis indicate that the null hypothesis is to be rejected.

Stem-and-leaf display A pictorial description of a set of data that combines features of the frequency distribution and the histogram.

Sum of squares The sum of the squared deviations of observations from their mean.

Summation sign A mathematical symbol that represents the sum of a set of numbers.

Symmetric distribution A distribution wherein one half is the mirror image of the other half.

***t* distributions** The statistical model used to test hypotheses about the mean of one population when the population standard deviation is *not* known, the differences between the means of two populations (independent or matched samples), the significance of certain correlation coefficients, and certain multiple comparison procedures in the analysis of variance.

***T* scores** Transformed scores that have a mean of 50 and standard deviation of 10.

Tetrachoric correlation coefficient A measure of the strength of the relationship between two dichotomous variables, used when the distributions underlying both variables are assumed to be normal.

Theoretical sampling distribution   An estimate of an experimental sampling distribution that is determined mathematically (rather than by drawing repeated samples).

Theta   In power analysis, an alternative hypothesis for the difference between two means that is *not* equal to zero and is expressed in raw units.

Transformed score   A score that has been altered mathematically to show its standing relative to a specified group.

Two-tailed test of significance   A statistical test wherein the null hypothesis may be rejected regardless of the direction of the results.

Type I error   Rejecting a null hypothesis that is actually true.

Type II error   Retaining a null hypothesis that is actually false.

Unimodal distribution   A distribution that has one pronounced peak when graphed as a frequency polygon.

Upper real limit   The upper end point of a class interval in a grouped frequency distribution.

Variability   The extent to which the scores in a specified group differ from one another, or how spread out or scattered the scores are.

Variable   Any characteristic that can take on different values.

Variance   A measure of the variability of the scores in a specified group, or how much the scores differ from one another; the square of the standard deviation.

Wilcoxon test   A nonparametric procedure for testing hypotheses about the difference between the locations of two matched populations, used when data are in the form of ranks.

Within-group variance   In analysis of variance, differences among the scores within groups. Reflects variation that cannot be explained by the experimental treatment(s) (error variance).

***Y*-intercept**   The point at which the regression line crosses the *Y*-axis.

***z* scores**   *Z* scores used in conjunction with the normal curve model.

***Z* scores**   Transformed scores that have a mean of 0 and standard deviation of 1.

Index